AWAKENING TO THE PLANT KINGDOM

By

Robert Shapiro and Julie Rapkin

.

Cassandra Press
San Rafael, Ca. 94915

**Cassandra Press
P.O. Box 868
San Rafael, Ca. 94915**

Printed in the United States of America.

First printing 1991

ISBN 0-945946-12-0

Library of Congress Catalogue Card Number 91-73455

Front cover art by Gary Hespenheide. Copyright © 1991 Cassandra Press.

Table Of Contents

Preface

In creating this sequel to *"Awakening to the Animal Kingdom,"* we felt it necessary to describe clearly the process by which we received this information. On several occasions we asked the plants to explain how they experienced the channeling process. The following was the clearest response we received:

"It is not normal for our species to see visual pictures. In this channeling process, it is possible for plants to see through the channeler's eyes. Since we do not have your senses, it is a new experience. We do not understand your language, so what occurs is an interpretation. It is similar to the situation of scanning a color into a computer which is then interpreted into mathematics on a graph. Within conversion from one language to another there are emotional subtleties that are lost. It is in the nature of print media that the inflections of the voice, which are attuned to the particular plant's mood, will be lost on paper.

"Through this being (Robert Shapiro), plants are able to transmit energy which is interpretable. Information is channeled through vibration pulse harmonics, which are converted into the language of English, utilizing the knowledge of the words in his brain."

We include this here so that you can appreciate our sense of wonder that such a process can occur and further the education of us all.

We would like to offer special thanks to Susan Matthews, Bruce Johnson, David Andorson, Sushila Vinyard and the rest of the Sedona gang, without whose assistance, this book would not have been created.

Introduction

I am the Spirit of the Grandfather trees. I begin now by saying that the plant kingdom, in all its varied shapes and forms that you view all over this beautiful planet, with few exceptions, did not originate here. The plant kingdom was brought here from other planetary systems. We are here in response to the needs of the human being. Many of our species cater to various animals' needs, and yet all of those animals are also here in response to the needs of the human being. Directly or indirectly, we are here for you.

There are a few exceptions to the idea that plants originated from elsewhere. I am referring to the continent that you now call Australia, both that which is showing and that which is lost. Parts of the lost continent of Australia will appear again at a later date, when they are needed, but for now I speak on the subject at hand. A few of the plants and many of the creatures of the animal kingdom that exist on the continent of Australia have originated on this planet. Australia is considered the "Experimental" continent.

Your scientists favor the theory that biological and microbiological life forms crawled out of the sea and populated the corners of the earth. It is partially true in the case of Australia, though it is populated by others, in addition to extensions of sea life. What has occurred on Australia is that the original "Seeders" of life on this planet induced the combining of the basic building blocks of life (including chemical reactions and interactions) with elements that were already in place (essentially earth, air, fire, and water). By that time, the sea was teaming with biological life, although not in as many forms as you now have (what you call the fishes). The one exception on your world that has experimental life forms that have genuinely evolved—as per the theories of science—is Australia.

Other life forms, kelp for example, have come from water planets and have been populated here and have re-populated themselves to maintain an aquatic life style. Kelp can provide nourishment, nutrients, and in some cases medical supplements that sustain the human being and other life forms that support the human being (insects, fishes, and so on).

I begin this book, which you will refer to as *Awakening to the Plant Kingdom*, with these few words to help you to see the overall plan which has been designed to maintain forms of sustenance to nurture and assist human beings. Plant forms can physically, spiritually, emotionally, and mentally nourish humankind. You will find many examples in later chapters. Our ability to live in harmony with each other, even though some species appear to be dominant and others more reticent, has allowed us to represent a harmonic example.

It is true that the more dominant species of plants are present as an example for you. You learn to see how dominance can work for you and sometimes against you. Those who cannot always defend themselves need encouragement and assistance. Plants help you to see which strong individuals among you, able to fend for themselves, can be called upon for the more demanding jobs (blaze new trails into the more rugged territories in your space, undersea, and underground explorations).

Know that the plant kingdom is ever handy to you. Though some species may come and go, we are ever in service to you, both through example and as sustenance. We will, to the best of our ability, remain your equal partners. We are here to sustain and nourish the variety of harmonious and multi-faceted evolving life forms to create the zoological and botanical wonder that is planet earth.

Chapter I

Bamboo

This is bamboo speaking. I am speaking diplomatically for the species as a whole. We have a united consciousness. I do not experience myself only as an individual personality. My orientation in your world is to provide not only a background for the theater of life but also to suggest how flexible life can be. We are very satisfied with what we are and receive pleasure through human interaction.

When you look at bamboo, notice where we grow and how we function as compartments. You can see that each segment is slightly different from the others, and yet we are perfect in our own right. The intention in our creation was to have a beautiful and practical plant involved in daily life. One can observe the beauty of our communication in sound and then harvest us and appreciate what we have to offer in the many accessories that can be created from our species. We are a support system in your lives, providing aid and comfort.

We are known to be of value through study and observation by philosophers and horticulturists. In our varied nature, we appreciate the flow of life as it unfolds. We are a portion of a larger picture. As a painter dips his brush into his pallet, we represent that pallet providing a backdrop to all life. We provide material for you to create from. We have no quarrel with being used for human purposes.

We are very intimate in our nature and communicate with each other through touch. If you are near a forest of bamboo, you will hear us creaking against each other. If there is a desire to communicate with a plant some distance away, the communication is passed through touch from plant to plant. We feel similar to you in this respect. True and lasting communication is gained through touch for your kind as well. Our natural inclination is to speak in this touching form.

Imagine a human being without bones in their body, as if they were on a stalk. As the wind blows through the human being, the arms and legs are gently curved to the arc of the wind. As a leaf falls and touches the body, the body reacts without pressure against the leaf. There is a flow, a dance, in constant motion associated with the world around it. This motion is interpreted to us in sound as well as in sight.

When a volume of air passes through a certain shape, depending upon the shape and speed of the air, different tones are created. The music will be the song of the motion of life. The Devas of the plant kingdom encourage our sounds, as they also communicate in music and song. They receive a great deal of pleasure from the sounds we create. We have been studied by certain cultures and religions. Those sensitively aware in our presence can see and hear the songs of the elementals associated with us. We influence, through song, the heart of humankind.

Our point of origin was on the planet Venus. We were created out of a desire for symbolically interacting with plant life and were synthesized scientifically. Variations of the bamboo family live all over your galaxy. The many varieties here on earth are representative of assorted forms elsewhere. We are here for beauty, communication, and to support you in furniture form. There is a tendency by some forces of creative business to view us as a weed. While we are appreciated by ancient and sophisticated cultures, there is still a tendency by modern humans to brush us aside. Monks in their civilizations and cultures understood that through our dance in the wind, we communicate not only our own nature but, through imitation, suggest new forms of communication in motion. We have more to offer and wish to remind you of our value.

We do not have vision, hearing, or taste as you sense them in your bodies. We have an increased energy zone with great sensitivity. Our auric field is so powerfully sensitive that we can intuit what the appearance of an object is. Our energy will reflect around it. You also have this capacity, although it is rarely used except in those with a loss of visual acuity or blindness. We have a sixth sense, that which feels and knows. We sense the proximity of someone or something. Our energy field wraps around their shape. If they are in motion, we feel how they move. If they make a sound, that sound interacts with our auric vibration.

When we procreate we come from the heart, rather than from some seasonal button which gets pushed. The reason that the human being has such difficulty in describing love is that it is primarily an emotional and physical description. Signs and symbols are much more demonstrative of love than intellectual descriptions. As you develop a greater understanding for the sacredness of yourselves, you will learn how to speak in the language of motion. This language has been pioneered by the Eastern philosophies, religions, and cultures through the martial and dance arts. The Western cultures, have been attempting to describe love metaphorically in intellectual terms, but have not developed an accepted standardized description. The identification, classification, and description of love does not allow for variation. The physical and emotional selves allow for unique individuality. Dance, in all of its forms, can develop signs and symbols as a direct expression from the heart.

The way to communicate with us is to move your hand gradually through the air, extending your energy field to sense the contact of the air with your auric field. In a meditation, imagine being in a field of bamboo, perhaps in a forest. Visualize the motion of one stalk as it gently rubs against other stalks. Let the sensation be that of the space around you expanding. Fill the room with your energy. For those of you who dance or do martial arts, move your arms and legs and imagine that your appendages are moving through something solid. Move slowly and feel the air as you pass through. If you suddenly move your hand through the air, you will feel a slight resistance. As you sensitize yourself, you will have a much greater capacity to communicate with us in person or through imagery. Send your energy out to us, as we do to you. This equal exchange is the best possible communion available at this time.

In your dream states you will not remember contact with us or with many other forms of life. What you do remember will be greatly colored by the events of your day. Those individuals that you hold as teachers will explain various life forms and introduce you to them so that you can communicate with them directly. You are being instructed in every moment. Some information may permeate your body or mental consciousness through the inspirational duct from your soul, which is known to you as the emotional self. This information may come to you suddenly and unexpectedly, as when you remember a dream in the middle of the day. It will be very fleeting if you do not act on it with your imagination.

We dream in our waking state. The motion that we experience is rhythmic and is sufficiently repetitive to be equated easily with chanting or meditation art forms. Through these motions and sounds, we experience the waking dream which allows us to experience other forms of life. When we are interacting with nature and when humans are not present, we experience universal life forms in front or in back of us. We feel their energetic presence and feel the benefits they have to offer the earth, and we interact on a personal level.

We experience space through touch. We have the capacity to experience linear time since we understand our actual growth cycle. We are able to tap into present time, which is the capacity of the individual to be totally focused in the moment. This is how we are able to access information that we as a species do not need to know, but desire to know. All knowledge exists in the intersection of linear time and now time.

Due to the longevity of the mineral kingdom, they are able to communicate with us in a much more lengthy fashion, to observe many species of bamboo coming and going over time. Since the life spans of rocks and minerals would be over millions of years, most of them communicate in an overview. Imagine how communication would change for an individual human being living millions of years.

In the nature of our interactions with the human being, we have been exposed to cultures which can trace their roots back thousands of years. This is necessary, not only to understand continuity but to appreciate and value it as well. The Western culture in the United States has the advantage of diversity, but has a disadvantage in that the governing bodies cannot trace their continuity back for more then a couple of hundred years. People tend to evolve geographically based on their interactions with the environment. We tend to reflect the cultures that we have been exposed to.

We have a variation of a mental body. We are able to access a wider margin of vocabulary than some other plants. We are an ancient plant in your civilization. We wish to stimulate a recognition within you of something deeper so that when you are in contact with us there is a sense of self-reflection. You have had time to consider the implication of interspecies communication, as well as the examples of nature as they flow into the human pattern. When you are around us, it is not unusual for you to feel reflective.

You do not yet fully grasp the language of a completely different species. For example, you have been attempting to create a form of verbal communication with dolphins for some time. The true nature of communication is not only through verbal, intellectual pursuit, but primarily through emotional states. As an enthusiastic child might scream with joy over a beautiful birthday cake, those screams would not be the same each time. Dolphins are much the same; while they have some repeated sounds, their sounds are based upon enthusiasm. When the human being attempts to communicate with us, we hear/feel/sense the entire experience of that human being: what they physically feel, as well as their unspoken subconscious and unconscious programs. This is why communications have not always been successful.

We have a multileveled communication to deal with. We send pulses to you and, the clearer the receptive medium or channel, the easier the communication. When you communicate with each other and experience the thrill of good news, you are having multileveled communication. We do not exactly understand your language on a mental level. Feel free to touch us and to use all of your senses to experience us. Tap lightly on the larger stalks and hear the resonance. Bamboo flutes can be made to communicate as well. When you use our body to produce tones for your own or others' amusement or pleasure, there is built-in communication. I suggest one of the more valuable ways to experience us is to listen to an experienced bamboo flute player hit the deeper tones of resonance. Relax into the tones and sense, feel, and imagine the air currents being slightly moved. When played in the proper sequence, this music can actually stimulate levels of physical and emotional action within the human being.

There are very few levels of the human emotional experience that we do not have the capacity to feel within ourselves. We feel love and hate. We can experience them through your communication and do not reject those feelings. It is our intention to emanate love, allowance, and curiosity and to stimulate a mirroring effect for humankind.

Begin to understand the need for new levels of communication that are simple to understand but not easy to put into practice. The greatest benefit to you will be a form of intimate telepathy that will be stimulated through feeling. Practice with your loved ones through touch or caring. Let your affection for a lover, daughter, or parent come through in physical touch and song. If song is difficult for you, attempt a musical instrument that moves the air. Begin to practice actual telepathic wave forms, which travel through all matter and for any distance without any loss in signal strength. Your levels of communication, which have become dependent upon technology, will be in jeopardy in the coming years.

You are here to develop all forms of loving communication, which sustains life. The challenges that you experience in the coming years will bring people together. You are encouraged to become universal citizens after you have recognized your earth citizenship. We resonate directly with the emotion of hope in the human being. Without hope, human beings have difficulty with growth. There is always the opportunity for change. Your race has been challenged to solve problems that seem insurmountable. You are able to achieve these goals in some form within the energy of hope, a building block of love. Allowance, tenderness, encouragement, fantasy, and the gift of productivity are energies that you feel around us.

If, by the turn of the century, there is not an emotional bond between human beings and the plants and animals that inhabit the earth, there may be the challenge of a point of no return. What would you do if the food that you eat in plant form suddenly began to die out? Begin to appreciate—in prayer, meditation, chanting, devotion, or by whatever means you are comfortable with—the earth that supports you. Allow and nurture all life and know that all have function and purpose here. When one species dies out because they are perceived of as not necessary or as annoying, other species will be affected. Learn to observe how nature functions and provides for all that exists. If harmony is to be achieved, the plant kingdom will need to feel the full impact of your resolution to become one with nature. Begin with your daily devotionals, in whatever form is of value to you. In order to experience and appreciate fully all that your planet has to offer, begin by falling in love with it.

In physical terms we are consumed, and we decay into the earth as does all physical matter. Portions of us on a physical and energy level will cycle through human beings. We understand the need of humans to harvest or burn us. We would appreciate it if you would utilize only a portion of a plant, so that we can grow again and provide for your needs. We appreciate a three-day notice, so that we may pass on our knowledge to our kind and say our farewell to physical life as we have known it. This allows us to prepare to move on to the next level of higher dimensional expression of ourselves. This sensation is like a spiral in constant motion.

While all plant life recognizes the point of origin for itself as light and love, after the completion of the life cycle on earth we identify the journey home with various stages of plant growth. We do not identify God in our own image exclusively. When a human being departs on their transition from life to death, they will often experience a form of their own bodies on the etheric level. We also experience a form of our youthful selves blooming or flowering as we move toward the afterlife. This gives our essence, or soul, an opportunity to explore the diversity that the universe has to offer.

Bamboo, as a food, is not fully appreciated in the West. You need to experiment with us in your own way and begin to use us in your gardens. We are a crop as well as a decorative plant. We feel no resentment in being eaten as long as we can be respected as a renewable resource. We recommend having bamboo or reed furniture in your homes as well. The subtle energy of our auric field does not die when we are harvested. When you sit on a bamboo bench, it reaches out and embraces you and encourages you to understand the nature of your being.

Chapter II

Banana

We are the bananas. We do not experience individual identities. Instead we have a group consciousness which connects us all. We represent the multiple levels of "social garb" that the human beings show to each other. In order to get to the heart and soul of the being, one must first peel off the outer layers. When the being is less developed, the outer layer is thick and hard. Other times, when the being is more centered and evolved in their life (more ripe so to speak), then the outer skin becomes soft and pliable and is much thinner. This is what we represent to your race.

We do not come from this planet, but have been adapted for life here. In our native place, in another star system called the Pleiades, we do not have a skin as you see on us here. We have a downy covering (not unlike the kiwi fruit) with a thin skin and a fuzzy appearance. The thick covering we have here on earth has been applied so that we could better survive the extreme heat we are exposed to. On the Pleiades there are not the extremes of temperature, but instead a constant temperate condition. The seeds within us are of a much lighter color, and our skin is almost white in appearance.

On earth, we must blend in with the rest of the plant kingdom and, as a result, we do have shades appropriate to your green plant color. Yellow is associated with the idea of the mental process. In order to get to the nourishment of our fruit, one must peel off the outer yellow layer. In terms of the human being, in order to get to the sweet inner portion, one must often peel off the outer layers of mental defensive systems.

In some systems, the color yellow is associated with the area of your body known as the solar plexus. This area of collection is a part of you that you draw strength from. We draw strength from our numbers. We grow in bunches, and it is in our nature to be familial. Your solar plexus energetically unites with other human beings to give and draw strength. When one of your numbers feels drained of energy from some contact, be it social, business, or personal, the actual loss of energy is through the solar plexus. This is perhaps why your ancient and more enlightened civilizations covered their solar plexus when they were working with individuals, either medically or philosophically. We are often invested with knowledge of your civilizations as it relates to us directly. The solar plexus is the portion of the human being that is most related to the banana.

In your society, it is unfashionable to be open and reveal yourself to those around you; yet in order to grow and assimilate into the true nature of all life, it is necessary to remind other forms of life—mineral, plant, animal, or ether—of your true personalities. This requires the opening of your heart and the releasing of the true personality of your race into your world. Sometimes you do this in prayer, other times in happiness and joy.

When you cause damage beyond your immediate needs, such as you do with pollution, you are trespassing against certain plant kingdoms. Those kingdoms are then unclear of your motivation, and they may find some way to let you know. If you have a type of fish or plant that is your food, and they can no longer be found, it is because that kingdom is expressing some form of effect for having been trespassed against. Other kingdoms have feelings, and if you shield your feelings from them, they will not know if an action you have taken was intentional or accidental. If you do not speak your feelings out loud, such as "People are starving and this is why we are using the soil so much!" we do not understand your actions.

Your Gods can Divine your thoughts, but as representatives of the plant kingdom, we communicate with your feelings. A farmer can go into a field and offer his prayers to God saying to the soil and seeds, "Please, we must have a good crop this year. I ask that your seeds be bountiful in your deliverance to us." The seeds can then feel inspired and understand your need. These ideas help you to realize the equality of all life. Plants can respond to you as you respond to the needs of your brothers and sisters.

We experience the Devic kingdom through fairies. They are very joyful and mischievous with the intention of bringing happiness and light into our lives and the lives of all that consume us. The banana is an unusual fruit. When you look at us, you are not sure what is inside until you peel back the skin and then, surprise, there is a present inside. Our fairies stay with us even after we are picked and distributed. Perhaps the slipping on the banana peel joke is related to the pranks of our fairies. Sometimes these funny thoughts are transmitted to humans because of the humorous nature of our Devic energy. Our interaction with the Devic kingdom can be seen as a green-gold light. This light energy is predominantly green with golden sparkles at its edges. When we grow in the wild and are unattended by human beings directly, our auric dance with the fairy kingdom is different from our dance during commercial production.

Bananas do dream, but not in the same way as you do. There is no separation from the body when we dream. You can tell when an individual banana is having visions because the fairies crowd around that particular banana. Certain bananas have more visions then others, and then the gold and green light swirls around and through them. The fairies dance and sing around those particular bananas. In our lore, the people who consume those bananas can experience wonderful dreams and great flashes of insight which lead to inventions that can benefit your race. Most of our visions are of home, our point of origin on the Pleiades. The origin is not so much a forest, but a plant system integrated within the society. When a human dreams after eating a banana, they may dream of this futuristic world.

We have instinctual feelings associated with procreation. We procreate reactively to the needs of the human being. When we feel the need, we are granted the Divine inspiration of the Creator, that which gives freely of all life, that which welcomes all life around us. Our connection to the Creator is deep. We sense no separation between ourselves and God, nor do we sense separation between you and anything in your world. All is of God, all is of the Creator, all is one. This is how we experience it.

You are a developing species which is adapting to this planet, and it is necessary for you to see God in your own image so that you feel safe, secure, and welcome. When a species is in its infancy, it perceives the Creator in its own image. This is not only native to the human being. When a plant species is new, or is cross-bred, its image of God is in its own image, so that it will not feel it is in a world that is foreign.

We have a strong sense of touch, and when we grow in bunches we become very familiar with each other. We experience space in relation to our touching ability. We do not experience time the way you do. Time does not evoke a sense of growth and learning for us; instead it is the cycle of life, from seed to fruit. We do not see or smell as you do. The senses that we do not have are made up for in the extra senses that our fairies have. They will do for us that which humans can do for themselves. They can be our ears, for example, if it is necessary for us to hear.

We have separate physical bodies, but we do not have individual mental bodies. If we are close enough to each other, our shared mental body connects to others of our own kind. Our spiritual bodies are associated with our Devic energy. When we know we are going to be planted, we send spiritual and emotional energy to that area. This energy is very powerful because of our interaction with the Devic kingdom. Residual emotional energy can still be found in an area where bananas were grown from 100 to 1,500 years ago. We prepare the ground in such a way that it is receptive to the return of the physical presence of the banana.

The banana is a spiritual food for humans. We can be used as a fragrant spiced tea, as a fried food, or eaten raw. Our auric field sprouts from our emotional body into our physical body and is present before we are planted and after we are eaten. Humans are learning how to use the auric field to their benefit. You can send your auric field energy to a plant or tree and address it from a distance. You might say, "Thank you for being here and supporting me." The adult human being may feel this is silly, but children, who know the value of life, will not.

If a human desires to receive inspiration from us, I recommend eating us at night before sleep time. When you consume the banana, realize that as you feel safe and loved, you also shed your outer layers and reveal more of yourself to your loved ones. The banana sheds its outer layers to offer itself as support in your lives. When you show your inner essence to your fellow human beings, know that you offer them sustenance simply from the essence of your presence. It is valuable to show others who you are, so they can benefit from the joy of that presence.

If you wish to share energies with us, imagine your auric field flooding around us and ask yourself to dream of our visions. Ask for what you want out loud. If you are feeling what you are saying, then you may be given a dream of banana life.

We do not communicate in the ways that you do. We communicate with each other through telepathy on a feeling level, not mental, but energetic. There may be a radiated communication associated with warmth or affection, in terms of emotional transference. We would not actually understand your language, but can understand your feelings.

We do not experience the concept of death in the way that you do. When the plant has to say good-bye to the fruit it has grown, there is some sadness and yet we are confident in yielding new crops. We understand that we are here to be harvested and eaten by human and animal life. The human being is still shielded from the knowledge of the process of life and death. When we are consumed, we do not feel any less alive. We blend into the fabric and structure of that which consumes us and we experience the energy of that being. Their auric field will be affected by the colors of our Devas. When we dissipate from that individual, we return to the geographical location of our next transformation or incarnation.

In the future, when you consume a banana, appreciate all the loving care and tender energy that our species has placed into it. Your Creator, in its infinite wisdom, has provided the banana plant for consumption by many kingdoms. Let those banana trees that grow wild be utilized by animals, and let your farmers occasionally encourage a tree to grow where humans will not consume the fruit. In this way, you can give something back to the land as an offering of appreciation for what has been given to you. When you take food from the land, it is offered freely. Learn to give back to the land, so that as you take you are also giving. Give with kindness and love, and in this way Mother Earth knows that you appreciate her. Value her as a role model and learn to give, to love, and to live.

Chapter III

Broccoli

Broccoli here. We are part of a larger family known as the cabbage family. We each have our own identity and personality. We represent cheerfulness and comedy. If ever a race of beings needed comedy, it is yours. You believe that you are here for some destiny which you perceive of as a burden. You are also trained to believe that you have an allegiance to various ideas. We are here in form and in function to remind you that life is not all struggle and striving for unattainable goals. Life can be silly, funny, and amusing.

The illusion that you experience has much to do with what you are taught and what you believe to be true. Since there is so much conflicting knowledge for you, you do not always know what to believe. We do not have that struggle since we do not carry knowledge the way you do. We are what we are, and we do not question it.

We originated on earth and were created by the Founders on this planet in preparation for the human being. We existed in the very distant past, died out for a time, and then returned. We exist for your consumption. In the past, the white inner core was used to make wine. We have an intimate arrangement with the mineral kingdom, and that is why we are appreciated as a good source for various minerals. We assimilate minerals very easily, and they enjoy their journey through us.

We do not have an emotional or instinctual relationship to procreation; instead our Devas and pixies sing to us and we know that it is time to re-create ourselves. The pixies are not unlike insects such as the butterfly, dragonfly, or praying mantis in appearance. Many of them have wings and, when sprayed with insecticides, must spend a long time recuperating. They each have a different song which helps other living things to grow and feel nourished.

Due to our cheerfulness, we have a very compatible relationship with our Deva and with the "little people" that exist beneath the surface of the earth. We communicate with the earth through song. The pixies sing in our dreams, and we experience a form of sound/dream that is musical. We hear/feel sound and when we have our sound dreams, we are in communication with those of our own kind near us. When we touch the earth, we have a greater sense of communion with ourselves.

We have pronounced physical, emotional, and spiritual bodies. We experience cheerfulness and well-being. We are very happy with what we are, and we enjoy a moist climate. We appreciate and give thanks for the radiation from the sun. When you consume us, eating the stem as well as the flower, you can experience this well-being with us.

We are aware of an energy field around our body. Our auric field is electromagnetic, and it radiates around us. It serves the function of a vehicle to assist in our materialization here. You also have that aspect in your auric body.

We have a composite identity which is different than a universal mind. It is a form of individual personality that is mixed together to create a generic personality. We do not have similar senses to your own. We use sound/touch as our primary mode of awareness for our surroundings. We experience the light and dark and infancy and maturity. We do not have a daily experience and do not notice the passing of time. We do have an apparatus that allows us to experience the present moment. I can utilize that aspect of the mental body that you refer to as thought when there is a strong enough desire from you for communication. The best way for humans to communicate with us in their garden is through song. We can feel/hear these songs and tunes. We prefer the human voice as it comes out of the vocal chords as compared to anything recorded. This way we can feel/understand you better.

We understand that we were created for consumption. We know when we are going to be harvested, even though the human beings around us do not often tell us. This is because of our compartmentalized awareness. When we experience our life, it is not related to time and space; instead it is like snapshots or freeze frame moments. We receive pictures of events before they actually happen and thus have an opportunity to prepare for harvesting.

Our essence will leave the plant about 24 hours after we are cut. The best way to consume us is out in the field; the second best is from a farmer's stand. The fresher the better. If we are eaten four to five days old, our nutrients will have paled somewhat. When you chew our bodies, we are not there. Our body is a gift that we give to you. We are aware that we are more than our form.

When our physical self dies, we do not experience much. When we are harvested and not grown again in the same location for three to seven years, we depart. If we are grown again, our consciousness comes into physical existence again. My understanding of the Creator is that it is a constant energy, which is the staff of life. We are given life when we are needed, when our presence is desired.

It is in the nature of all life to be fed and supported by other life; we do not appreciate being fed toxic insecticides. We feel that insecticides separate us from natural life here. In time you will embrace the earth and live here as the natural being that you are. Right now you seem to be greatly enamored by technology, which you believe makes your life better. You are living and eating and breathing technology—is it worth it? There is a degree of unconsciousness that keeps you from recognizing yourself in all other things. When you gain a greater respect for your own physical bodies, you will gain a greater respect for ours.

A viable alternative to the use of pesticides in your industrial civilization would be the hiring of people (plant priests) who are in touch with the plant kingdom. They would be permitted to walk along farms everywhere and talk to the birds, the soil, the trees, and the insects. They could communicate to the animals and insects what is needed and desired by humans: "Please do not eat the crops." The plant priests could make an offering to the animals and insects. Even organic fertilizers can be overused. Organic fertilizers that do not naturally occur in the region feel uncomfortable to us, but not as much as the chemical fertilizers do.

We appreciate it when you recognize that the food you plant in the earth to feed yourselves exists here for you, regardless of how much tending of the soil is needed or watering or praying for rain. These things are offered to you because the Creative force understands your need to be nourished and cherished. We see that you do not understand the great value of your own physical bodies. Understand that you are created to nourish the Creator itself. We, as plants, become food. You are not consumed, but you are created by that Creative force as its loving children. Can you not cherish and love yourselves with that understanding?

Chapter IV

Catnip

We are catnip. I am speaking to you as an individual plant, though I do not think of myself as a named individual as you do. We are all equal, although we all do seem to be individuals. We believe that our symbolical focus in your lives has to do with the rekindling of romance; our archetypal significance is to help instigate the adventurous soul and to expand the magnanimous heart. We consider ourselves to be brewers of good cheer.

We got our name because cats discovered us and would eat us and act wild and crazy. We were not originally intended to be used internally, but were actually designed to be used in poultices and external rubs for human beings as well as for horses and other creatures. I am here to request that those who prepare herbal remedies look again at catnip with favor as an external application. There are those who make a tea of catnip and drink it. It's not really very tasty, and if you drink too much, you will feel dreamy and have difficulty focusing your clear thought. It might stimulate erratic dreams. Although there may be some uses of catnip in tinctures, I only know I came through to encourage you to explore external uses more. Much of the knowledge you need is just waiting for somebody to seek it out.

We can be grown in pots indoors in houses and appreciated in that way. To enhance your adventurer and romantic spirits, you might rub the leaves a little bit on the inside of your wrists while thinking thoughts of imagination and adventure. We can be used effectively either dried or fresh. If you wish to pick our leaves, we ask that you give us a couple of days notice, perhaps saying something like, "I am going to ask you for a few leaves in a few days. Please prepare yourself." This gives us an opportunity to communicate our knowledge to each other in a kind of tribal fashion. Otherwise, we understand that we were created to be used by humans and do not mind being used.

If there is sensitivity to the skin, then smelling catnip can be effective. If the catnip is dried and burned, the smell can be as effective as touching us. Just breathe it in gently; taking a deep breath would be too much. Just use us occasionally; we're not really incense. Perhaps if there was someone coming that you were romantically interested in, you could burn a little catnip along with some dried mint.

Tapping the soul consciousness of catnip, I seem to have been created on this planet, perhaps in the time of Atlantis (I'm not really certain, but it was a highly technological society), in a huge underground laboratory designed especially to create plant species that would either have a use in that present time or one in the future. Although we do not have senses in terms of sight or sound, I find I can use this medium's sense and tap the sources that he possesses. The laboratory was huge, at least 3,000 feet wide and 5,000 feet long, and maybe 12 feet high. Outside a huge window was the sea (which might also have been an experimental habitat as well) with beautiful, brightly colored sea creatures similar to your goldfish and sunfish, but as large as dolphins and whales of today. We were underground, perhaps on the coast of some continent or on an island.

I am unclear as to who our creators were. Apparently the society that was raising all these plants was very sensitive to pollen; just a tiny particle could cause serious injury, even death. Therefore, everything was fully automated and completely run by robots. The robots looked somewhat humanoid with long faces, and they were odd looking, somewhat ungainly, and not very attractive. The whole laboratory seemed to be a kind of magnanimous gesture on the part of these beings simply to provide plant species for other races, other creatures, animals, and future humans; it does not seem that we were being created for them.

There were thousands of plants grown there. I believe quite a few are still in existence today, mostly in the category of spices or herbs that can be used in healing remedies for animals or people, as well as cooking herbs and spices. There seems to be a vast array of different types. Some of the plants that you have now that you think of as weeds are either left over from a time in the past when they were very beneficial or are waiting to have their uses discovered in the future. There are no plants that are not valuable.

There were many varieties of catnip there, so many I cannot really think of them all (I don't really "think"). Today catnip plants are mostly variations of green and brown, but in the laboratory all the plants grew in a wild panorama of color, including colors you've never seen in plants before. In each species there were many different colors. In the catnip species alone, every color of the rainbow, every color of an artist's palette would be found there. Perhaps my favorite one is purple with pink tips on the fronds and leaves.

We seem to share our Devic energy with about 15 other species, from big trees to some that look like mold or spurs. Therefore, our

Deva is not always available to us; our Deva is often occupied elsewhere. Our Devic energy is definitely something greater than ourselves.

We use our auric field to form up physically, like a canvas or a media that allows us to create our physical appearance. We seem to use the density of our auric field to communicate with one another. We do not seem to communicate with many other species except dolphins (perhaps from our connection in the ancient laboratory by the sea). Our knowledge of humankind has come from our communication with the dolphins.

We do seem to have four bodies, although some are more developed. Our spiritual bodies seem to be at least as well-developed as your own. Our mental bodies seem to need more development. We also want the veils to be removed so we can understand ourselves. Emotionally, we do seem to be happy most of the time, always cheerful. We have a strong sense of continuity, which contributes to our cheerfulness.

If humans wish to communicate with us, perhaps emotionally would be the best way since our emotional bodies seem to be well-developed. If you imagine a sense of continuity of past, present, and future and find a reason to be cheerful, you might feel our presence more. We could assist you in having pleasant dreams. This is one of the reasons I encourage bringing catnip indoors as a house plant.

We often dream somewhat similarly to humans, Since all life on this planet seeks its origin and attempts to understand its beginning, we always dream of the past. Also not unlike you, we are veiled from knowing more about ourselves. This veil seems stuck, permanent; I have not been able to break beyond it. Perhaps this is something we have in common with humans: the veil to understanding ourselves.

I have no idea of our future. I am of the idea that we will be used much more in your immediate future, somehow having to do with agriculture. I am seeing us on the surface of the ground, having been grown and placed underneath rolling plants, but I am unclear as to our purpose.

Since genetics is becoming popular on your planet, I want to say that a very toxic version of ourselves could accidentally be made. In the past, as different colors were experimented with, the third generation offspring began to develop a toxicity which continued to multiply until it became quite dangerous. The toxicity was so extreme that it damaged some of the robots. Therefore, the green and brown varieties are the only ones who survived and made it to the earth's surface. I do not recommend bringing out the colors again.

Death to us is simply change; we do not feel a sense of cessation of life. We seem to immediately go back to the laboratory. We don't remember any more, but that is our experience of the change of life.

We feel that humans and catnip are connected somehow on the soul level, perhaps from this ancient civilization. During this time of change for our planet that we are living on, all species will discover more about themselves and each other. We feel that you will be using us in some new and revolutionary way for yourselves and even creating some new kind of communication with us and other members of the plant kingdom as well. We salute you in your constant search for self-discovery and wish you well on your journey.

Chapter V

Cholla

Cholla here. I am what you might refer to as a desert plant of the southwest, perhaps of the cactus family. You would recognize us by our arms that reach out with prickles or needles on them. We represent to you the embrace of Mother Earth, her good side as well as her strict side. When Mother Earth offers all she has for you, she offers not only the comfortable and benign but also the challenging and sometimes painful.

As far as I am able to tell, we originated on the planet with the rings, I believe you call it Saturn, in a laboratory where plant species were experimented with for use throughout your solar system. We seem to have been synthesized from other species in this laboratory inside the planet of Saturn. The needles were originally from a larger plant, sort of a tree with no leaves, and were added to us. It appears that we were designed originally to keep wild animals away from other more sensitive or fragile plants rather than to protect ourselves personally. I believe that the energy of Saturn at that time would be more aligned with the fifth dimension rather than your current third, in the sense that the speed of light was much expanded which caused all energies to move at 10 X 10 times your speed of light. This is, I believe, the basic formula for the difference in dimensions; if you move from the third to the fourth dimension, for example, you will experience that the speed of light is 10 times faster.

We also seem to exist in another form inside the earth in some subterranean civilization where we are an edible, succulent, sort of fleshy plant which has no needles. Perhaps your agricultural people could cross us with other succulent plants and create something edible without needles, as was done in this other civilization.

We seem to represent to you the need to cooperate and embrace your planet and one another. In the past it seemed possible to do this only very conditionally (because of needles or prickles). As you learn more and more how to cooperate with each other, with the planet, and with the animal, plant, and mineral kingdoms in order to live holistically and harmoniously, you may wish to remove, at least symbolically, the needles or prickles from your arms so there is not so much of a conditional interaction with other species or other humans. In the coming years, there will be many situations that will require cooperation between the races of men and women, such as pollution, so it will no longer be convenient or beneficial to continue to have these conditions keep you apart from one another. This may not happen immediately, but consider the idea that it is not only more pleasant but certainly in the long run will be essential in order to create a strong united front to take on problems of pollution and ecology, such as the hole in the ozone layer.

Although your scientists are working on devices using your space programs to replenish the ozone layer, they will find that nothing will work as well as the replanting of Mother Earth. If you wish to get a headstart, begin planting, greening the earth—a garden to grow food, flowers, trees, even cactuses. There are forms of technology available to green over the deserts. If you really want to replenish the ozone layer, the best thing to do is to plant, plant, plant.

We have an interesting relationship with our Deva. Since we were synthesized rather than naturally evolved, we seem to have sort of one and a half Devas. In other words, we have our primary Deva, then we have a sort of "child" Deva that represents the prickles that were added to us. Our primary Devic energy is very powerfully feminine—very beneficial, very loving, and very amusing. The "child" Deva is more masculine and somewhat apprehensive, shy, and vaguely afraid. Our primary Devic energy spends a lot of time embracing and nurturing and loving this secondary Devic energy. It's almost as if this secondary Devic energy is apprenticed to the primary Devic energy, almost a parent/child relationship where the new Devic energy is being nurtured, born and raised, as it were, and at the same time the primary Devic energy is learning about needs that have more to do with fear than harmony. We plants sometimes feel pulled one way or the other, sort of like having a split personality.

We feel a very strong sense of family rather than individuality, but it's more like our entire species is our family, rather than your type of family, where there are just a few people.

Our dream life is very interesting and vigorous. We have dreams that may last for months at a time. People who study us have noticed that sometimes our plants will die off for no apparent reason, seemingly as if they had been unattended. These plants were actively involved in their dream life and basically abandoned their physical apparatus.

When we dream, we are not drawn back to Saturn as one might expect. Instead, we are invariably drawn inside planet earth where there are many different deep caverns and where one can explore and find different plants and occasional animal species that have managed to survive from ancient times. Our energy is pulled deep within the planet where we explore as a cave explorer might, all the nooks and crannies and underground wonders that might be found there. This is speculation on my part, but it seems as if we are evolving to become more like our underground species, and perhaps this is why we are drawn there so strongly, almost like a magnet. I supposed you might say we are dreaming about our future selves. I have noticed that human beings can spend too much time dreaming or daydreaming and neglect their bodies as well.

We do not experience senses like human beings do. This form of communication now, this mediumship, is new to me, which is perhaps why my speech is slow and halting. Our major contact with human beings is in our dream time when humans are also in their dream time, not their deep dream time but just as they fall asleep or begin to wake up. Even with this medium, I seem to be communicating energetically along the dream line of energy. I'm sure many of you have seen us or at least a form of us in your dreams; you might have thought you were on another planet when you saw us! If you wish to communicate with us, you might look at a picture of us before you go to sleep, and then have a fun dream with us; anything is possible in dream land.

We communicate with the other kingdoms in our dreams, as well, although we experience some physical contact with our neighbors that grow near us, with the insects and animals that touch us, and with the minerals in the soil.

We have a certain sense of emotions, but more as an auric field or light body, which we use as a sensing device not unlike your own electronic sensing devices in that we radiate our energy out. Whatever it happens to wrap around or bounce off of we will understand roughly the shape, size, and perhaps density or texture of, so that we can orient ourselves in our world.

Another unusual aspect of cholla is that I am not aware of us having any basic drive to stay alive or to perpetuate the species. It is not that we do not have a will to live; it is that we are not attached to living in our current state of being. We are constantly envisioning our future selves and the way we will look in the future without these prickles and with our primary and secondary Devic energies combined into one, so we do not have a overwhelming drive to continue the species as it is. This is part of the reason we are dying out as an apparent species on the surface of your planet now. We will probably disappear from the planet, except for rare locations in the wild, for perhaps a million years or so, and will reappear when our next form shows itself.

Our mental body is functioning along the lines of a universal translator which translates our energetic field, our auric field, which functions very fast, faster than most of your computers could record it even if your computers were set on the maximum input potential. We have a sort of mental body which has the capacity to translate our energy into thought for you or perhaps some other species that would require that.

In terms of a spiritual body, we definitely feel a strong sense of divinity in terms of our Divine interaction with Spirit Mother Earth. We sing with Mother Earth's auric field, and we feel like Mother Earth has adopted us and taken us into her arms. We feel a strong sense of our spiritual bodies, very strong, in very specific alignment with Mother Earth as an entity or being.

We experience the Creator—that which creates, simulates, and allows all life to change, grow, and expand on its own—we experience that Being, or that sum total of Being, as a friend, as a kind of cohort (I'm searching your vocabulary). We experience the Creator as a kind of friendly counselor, someone/something that is loving and at the same time guiding us, almost with an "invisible hand," as your writer has called it. Although this hand is moving us toward some destiny that we are uncertain of, because we trust the Creator, we allow ourselves to be moved, almost as sheep would allow themselves to be moved by the shepherd because of the built-in trust and sense of alliance. This is my best description of our relationship to the Creator. I am not referring to our Creator in the factory on Saturn; we understand that they were only doing the Creator's bidding.

We have felt unappreciated and unneeded by humans. For a time, this caused us to feel somewhat unhappy, but we did not understand that it is a part of our declining cycle that will allow us eventually to reappear again—a million years or so down the road—in our new, more harmonious form. We now recognize that the human being's needs are actually pouring into our evolutionary cycle. It is as if our philosophy has changed in recent years to the belief that the Creator is allowing such a thing to take place because it will speed our evolution. When most of us are gone from the surface of the planet, we will only be gone physically but will continue in dream time on the spiritual and emotional planes. However, we would prefer to continue with our gradual change in evolution. Without humankind messing with us, it would probably take 10 to 20 thousand years for us to gradually die out. We do feel rushed. And we do wish you could stop long enough in your land clearing to admire us and appreciate our beauty and diversity.

I believe that the source of the apprehensiveness of our secondary Devic energy is associated with the prickles, with the knowledge that contact by other species will cause them physical pain. I believe it was the intention of the Creator to surround humans and their world with symbolic representations in the plant, animal, and minerals kingdom of facets of the human character as well as what your life is like or could be like. I believe our Devic energy is not happy about being prickly and is looking forward to the release of the prickles so it can feel the joy it naturally contains. Right now, it knows that the only way it can contact human beings is through a human being's discomfort, and this the Devic energy finds really unsatisfactory.

It seems that human beings find themselves at a crossroads, that human beings appear to be evolving as well as we are. It seems that the next stages of your evolution have to do with your ability to cooperate with yourself and parts of yourself and to find enough love in your heart to appreciate the parts of yourself that are in pain or suffering, so there seems to be some symbolic similarity to our Devic energy. It seems that those stages of life that you are in now are very challenging and difficult but will prepare you for a larger philosophy of life. Most of us plants and animals and minerals are very specialized. While humans have developed specialties, you also have an ability that most plants and animals find difficult, and that is your ability to be versatile.

We have learned to value our species. We have had to face the fact that we have needles and prickles that have inflicted discomfort on others. We have had to accept it and honor it and allow it. A few people come to study our beauty, but most times it's, "Ouch! What was that? That didn't feel so good."

As a species, you are evolving in a beneficial way. In the past, perhaps species such as my own have forced you to become aware of your surroundings so you would find ways to cooperate with it, to become compatible with it, to create a strong united front. Perhaps when you do, there will be less need for plants with prickles, and then we will be allowed to release our needles. As you become more aware, we can then return to our native appearance.

Chapter VI

Corn

I am the plant you refer to as corn. I speak to you as the Deva associated with our super-consciousness, the Corn Goddess. I have a very strong spiritual body that not only connects me cosmically to my point of origin but also embraces the earth as a nurturing parent.

Goddess energy represents the heart, the soul, and the mother of all. Before it interacts with the human race, corn energy sources its strength through the planet Venus. I came from an ancient civilization which existed on Venus. Venus is the heart and mother of your solar system. Those who brought us to the earth were aware of our spiritual significance, as well as our ability to sustain physical life.

Your souls came from other points of origin. You may feel at times that you would rather be elsewhere. You may also feel at times bound to earth as a prisoner might feel bound to a prison. If you eat more corn, you find that my interaction with your emotional and spiritual bodies helps you to feel more at home here. The spiritual properties that I possess have not been completely discovered by science, as it is practiced in your Western world. First the emotional body, while being sustained through the consumption of corn, acts as a conduit to the spiritual body. Then I assist your spiritual body in its unification with the earth.

I understand that humankind is in school here to re-learn what they naturally know at birth and begin to remember in old age. Life is holy and given for the purpose of inspiring wonder and joy. Humans are attempting to regain this knowledge. Corn and other plants understand that you are questing for this goal.

I am here to inspire you and to enhance your ancient tales. Many Native American stories weave us into the history of life as the Corn Goddess. The group of people that you refer to as "Indians" have been, for some time, pioneer beings. When a new life form is seeded upon a given planet, many will travel in groups, hence the idea of tribes. They will populate the planets by creating and by being created as the prototype physical life form of that planet. In that sense, they bring the food that sustains them the most.

Corn sustains the Native American's spiritual, emotional, mental, and physical bodies, as represented by the directions and the seasons. I represent the herald of the seasons. As any farmer knows, to watch stalks of corn in the field as they grow and mature is a wondrous sight. I am planted in one season, and when the time comes to gather in the mature crop of corn, there is the celebration of another season.

In some populations, I am the basic food and I am often used in healing rituals. Most of the medicinal properties of the corn kernels and the cob itself have been discovered by Native American tribes. There are still some uses of the corn's root which have not yet been explored. In very small quantities, red corn is utilized for its aphrodisiac qualities. Eaten in excess, it causes nervousness. (I recommend consulting a good herbalist for correct proportions). Corn, boiled with anise seed and small quantities of dandelion root and made into a weak tea, helps one to experience soothing dreams.

I do not dream exactly as you do. I understand that you experience your dreams as a comparative state of consciousness to your conscious awake state. I have a meditative state that allows me to be in an altered state, which is only slightly modified from our normal state of being. It is not unlike your daydreams. It is a way to connect with our home planet, which is more the unity of the corn consciousness rather than a geographic location.

When I am planted in a field, I choose to feel every row and plant as a unit. Therefore, I do not experience spatial references in the same way as you; I feel myself as the entire field of plants. When an area of corn is accidentally destroyed, some element of protest is expressed from the rest of the corn. Corn shrivels a bit or makes a sound that only the observant farmer notices. I do not shrink out of fear; instead I understand that I am here to sustain you. I am prepared, at any moment, to offer myself in support of my true purpose on this planet.

I do not actually die. The spirit of corn lives on. While my physical presence can be removed, as long as I am needed on this planet I have an energy presence here. This is why those who are spiritual people very frequently plant us in precisely the same patterns, same rows or fields, as we have been planted before. They feel our spirit presence so strongly there. I do not return, as do your souls, to some source, since I am needed here. My presence helps to bind and balm your presence with Mother Earth.

I have a sense of touch similar to that of the human being. I know when I am touched, and I am aware when someone or something is near that is not of my own kind. I have a strong energy field that radiates with an awareness of up to six feet. I have an ability to respond to the change in weather conditions and to the changes from day to night.

Unlike yourself, I do not collapse my energy field, even with the approach of inclement weather or an individual or machine to harvest me. I do not feel pain in the same way you do, but there is something akin to discomfort. When I am cut, bruised, or damaged, without an intent to be consumed or harvested and without an appreciation for what I offer, my feelings are disappointed. My purpose for that particular crop will be lost when we are sacrificed due to humans' inability to understand their own creations.

Diseases that affect a crop can be assisted through meditation or prayerful communications. Earth must provide nourishment for all, not just a specific few; this is why birds must occasionally invade the crops. They themselves sustain other creatures, even humans. The diseases which wipe out a crop provide for the insect kingdom. Reverential offerings to Mother Earth in time provide fields given directly to you.

I am not attached to how you utilize the corn after it has been harvested. I hope you realize that your intent has a great and powerful effect on everything that you eat. If you intend that every food you eat sustains your well being, if you pray that it does so and thank it, then it definitely will sustain you. Foods can be greatly improved upon in this manner.

I prefer that natural pesticides be used by your farmers. I understand that in your present day, you have become so removed from an understanding and connection to all life that you feel out of control. You do not yet understand the word balance. In the past, native people's dances, prayers, and offerings were equal to the cyclical nature of life. What they offered and asked for was equal to what they gave. In our life as well, everything gives and takes equally. The fly consumes what it must and is consumed by other creatures. The cycle of life is in constant motion.

Human diseases and wars, which often end in great loss of life, are methods of sacrificing or giving back in some way. This is why many ancient religions performed human or animal sacrifices. Recognize that their intention was to make an offering so that nature would give back to them in return. They did not realize it was not necessary to sacrifice a life, but they perceived it as the way of nature.

Now humans can offer different and greater gifts, such as prayers of thankfulness to Mother Earth. For city dwellers, performing a kindness to another human being, animal, or plant is giving as well as receiving. If enough human beings perform these duties in creative ways, they will cut down on war and famine due simply to their willingness to give.

You can best communicate with us when you feel a consuming love for the earth and mean it. I understand that you do not fully grasp the power of your emotional body and emotional messages. Miscommunication happens when your word is not equal to your emotional intent. When children wish to communicate with us, they can be cheerful and laughing. Children are the beings in your species who can hear the laughter of the corn Devas. I feel, in my natural state of being, joy for the gift of life.

I have a very powerful emotional body. I use the auric field of my emotional body as a single identity and communicate through radiant energy. This helps me to bind myself with others of my type. I feel a kinship with all corn. Corn shares a mass consciousness through thoughts, memories, experiences, observations, and all else that any intelligent species absorbs.

All plant kingdoms have accumulated knowledge. We enjoy exchanging this information but do not feel a need to do so. I do not experience a mental body in terms of linear thought. While I can communicate that which is apparent, I experience a more vertical thought. Whatever I need to know I know at once; it is retained permanently and is used when needed. It is not stored in a conscious or subconscious mind. This knowing is in direct conscious response to the stimulation of the soul. Corn can encourage imagination, especially when eaten fresh, right off of the stalk.

I hope that you will allow yourselves to experience more of the harmony of all life on this planet. Since the age of science and technology, the sanctity of all life forms has been seriously neglected by humans. All life forms work synchronistically with each other to sustain each other's needs and life styles. I hope that in time you will all become increasingly aware of our interconnectedness and appreciate not only the wonder of it but the holiness of it as well.

The life of any species is due in large part to the song that it sings. In life, a human being is in song at all times. The song is strong in youth and weak and tender in old age. To prepare for the transition, the song must become quieter to hear the song of the Creator and follow the music home. All life sings.

Chapter VII

Cottonwood Tree

We are the cottonwood trees. We are aware of our geographical location in a mountainous region, close to a stream in northern Arizona. We do not have egocentric, individual identities; instead we feel a sense of belonging and family consciousness. That is the best way we can describe it. We do not have dominant individuals; we are all equal and affectionate with each other.

We represent the graceful entry into transitions. Much of what we do is transitional. Our seeding process for the next generation is done in such a way that portions of us must be involved in metamorphosis. Similarly, when a woman gives birth, the placenta, which supports a new being, is released into the world, then dies. Our transformative nature is designed to help human beings understand their own abilities to transform energies, substances, and experiences into something that seeds new generations of ideas, thoughts, beings, and life forms.

The idea of continuity has been conveyed by your religions and less successfully by your philosophies, which tend to be politicized. Your religions have had limited success in bringing the idea of continuity of worthwhile ideas and experiences into the next generation. Since technology has come along, many difficult new problems have developed. Many of your religions know not how to answer these problems, so they grasp at straws and give conventional answers that were once useful and valuable.

Now people are wondering: What about continuity? What about the next generation? We see when we drop our seeds and release our leaves that that which cannot be accomplished by the parent tree may be accomplished by its offspring. This is something we believe in absolutely. The reason we discuss this at such length is because it is important for you to know that, even though your problems look so critical and difficult to solve, the new generations will have many of the tools and abilities necessary to solve them. We are not suggesting that you give up trying to solve these challenges. We encourage you to understand that for your existence to continue there needs to be trust and hope. Do not lose hope, my friends, there are opportunities for the continuity of your civilizations.

Do not just assume that because the youngsters look unusual or have strange habits, dances, or musical tastes they will not evolve into great contributors to your society. Remember how you were once considered odd and unusual by your parents in terms of your likes and dislikes? Be aware that, even though these days children are exposed to extremes that you prefer they were not, they seem to grow up before their time. This impacts upon them the urgency of life's problems and how they must be solved and how the old way, that which you have created with your technological society, is not sufficient to solve them. Many of them have the desire to go back to nature and seek out examples. They will begin to imitate life as it exists in the true idea of Mother Nature.

We originated in the dawn of time beneath the earth and do not exist in this form anyplace else. There is a laboratory which still exists inside the earth—some people say it is run by the gods. It is run by people with bodies unlike your own; however, their bodies do not decay. Their environment is very carefully controlled, and they are timeless. We refer to them as the "Eternals." They have done the Creator's bidding well through simulation of many different plant and animal species, using the elements found here on this planet. They created plants that would survive well within the mineral and chemical environment on this planet.

The story of the friendship with minerals is passed down through generations. When we were first planted in the earth, we were grown in water in a fashion similar to hydroponics. When we were placed in the soil, we were told by the Eternals to expect the unexpected. We had a sudden change of physical environment. When roots grow in soil there is a degree of resistance, and in the beginning we had to communicate with the soil. We asked, "Why do you resist our presence?" It answered, "It is not that we resist your presence. It is that we welcome you in the best way that we know how, through physical touch. We must remain in constant close touch with you. It is our job to encourage your roots to become strong and viable on this planet, for only that which is strong will live forever." We understood this and soon began to notice that we were taking in very minute particles of the minerals in the soil which had been embracing us so tightly. We find minerals inside us now, not only embracing us from the outside, but from the inside as well. The minerals said, "We have been here much longer than you, awaiting this means of communion. We are pleased not only to make your acquaintance from the outside, but to become an intimate portion of you. We hope that this introduction to intimacy allows you to appreciate the touch you experience here on earth. Flying creatures, land animals, and humans will touch you. As we embrace you from without as well as from within, we welcome you to this planet and say happy earth time."

Our communication with the mineral kingdom has developed out of necessity. When we were originally placed here, we did not grasp that minerals were more than friends. We did not realize that they were family and were vital to our survival. We communicate with them as they pass through us through feelings and touch.

Since you are part of the mineral kingdom, we communicate with you best through touch as well. While you can always use your imagination to speak with us, especially children, we most appreciate gentle touch and song. This is not to suggest that we like loud stereos being placed under our trees. When you are exposed to us, please turn down the volume on the radio, as we feel it as a disharmonious experience. We very much enjoy song that is generated by human beings or their instruments.

We feel a sense of close bonding with the Devas. Many of them sing from our branches. They are musically oriented and cheerful and we enjoy the pleasure of their company. When you are standing near a large old cottonwood tree, you may unexpectedly see a branch move with no wind or creature jumping on that branch. If you break into song, you will delight the Devas there. The Devas share their funny experiences with us and don't miss very much. We share a pleasure of storytelling which incorporates humor and childlike delights. Our older members, eternal spiritual trees, are fond of speaking of the old days and how our current evolution has resolved our difficulties and challenges.

We offer beauty, advice, and encouragement to you in the perpetual continuity of life. If an individual tree is cut down or struck by lightning, we all mourn its loss, as you would mourn the loss of a family member. Through our grief we gain strength and resolution. Even though plants die out, they return in a new and unique form. Never give up your faith that this is so.

We have a form of conscious dreaming similar to your daydreams, although ours are communal. We see ourselves planted together in a circle or semicircle. We are very lightly moved by the wind which causes our leaves and branches to touch each other.

We have a sense of touch, but not of vision. We have a mind's eye and mental perception. We do not hear, but we feel. This is not unlike a deaf person in your world who senses the surroundings and who is aware of the vibration of music. Music moves air in a specific way which, in a sense, offers us a form of hearing.

We experience time in cycles of day and night. This awareness fosters a belief in continuity and in the value of time. Time is a very valuable tool which allows an investment in the resolution of certain problems. Please don't be discouraged by repetitions of the same problem. You may be resolving another soul's problem through communion with them. A soul can share what the physical personality has accomplished in problem solving.

We value time and space equally, but our spatial reference is different from your own. What kind of life would it be for a tree such as ourselves, which spreads itself out vertically and horizontally, if it were cramped into a small space? Space is essential to us for growth and movement and for becoming all that we are. Human beings, while they often gather in small spaces to work together, need to give themselves time to walk in open spaces to appreciate nature. This can refresh your knowledge of the openness and expansiveness of space.

We are aware of our auric field and know that it radiates out, way beyond the perimeter of our physical self. Only in times of crisis does it pull back within us. Through this field we can experience life at a distance, such as when a deer is approaching or when a pair of hikers are moving through the woods.

We communicate primarily through the auric field pulse. This is common among plant species. We communicate through group storytelling and a tonal structure that runs beneath the earth. The sound we identify with the inner earth gives us a sense of communion. The sound is not unlike your wire from telephone to telephone. It is an auric vibratory connection that utilizes the tone of the earth as a means to touch.

We do not actually understand your language; however, we understand that the universal language is converted into what we experience. A beat frequency which has many variables is the element of the universal language that translates anything and everything. Your scientists will eventually create a universal translator through the understanding of music, tone, vibration, and beat frequencies.

Our mental body is different than yours. We call it the secondary self. It allows us to understand the world in which we live, while at the same time it creates a link to ourselves in the past and future. Our mental body has the ability to recognize and appreciate, to value and consider. We understand the evolution of creation. We see the mental body as being sacred and close to God. Through that level of recognition, we harmonize our mental bodies with our emotional and spiritual bodies. Perhaps your struggle is different, and we want to encourage you to know that Mother Earth has never bitten off more than she can chew. She is always willing to work with you.

Humankind has separate bodies integrated into one unit. In our case, we feel a sense of harmony, with no disharmonious element struggling for attention. Much of what we came here to do has been resolved. Our spiritual, physical, and emotional bodies operate in complete harmony.

We feel joy as a constant state. We also feel pain when we are cut down. This is why we appreciate being asked—before you cut us down. We appreciate being informed a few days before the trees are cut down. Please state what is going to happen in the kindest way possible, not unlike the way a doctor speaks with a child before an operation. Imagine that you are talking to a child, without the childlike words. If we know that you are not taking this action out of anger or resentment, it makes it easier for us. When you experience an injury, it is painful and no one enjoys it. It is so much more painful when you experience an injury that is done out of malice from another individual. Then the injury is not only skin deep but is emotional with lingering pain.

In order to prepare to die, we like a few days to concentrate our experience and to tell any last stories that we may wish to pass on to our fellow trees and the Devas that frequently sit on our branches. Then we sing a very specific tone that, if you could hear it, sounds wavering and announces the end of our life. It embraces the next level of existence. It exists vertically from our roots within the earth, upward into the sky and the unknown. We make this wavering sound for 14 hours at the most, and this helps us to evolve in our essence. We go into the auric body of the earth, where we are embraced and experience what we know to be the Creator. We follow the fundamentals of earth love and the orientation of earth spirit to its final destiny. We move in the earth's dancing aura, magnetically and electrically. From there we move on to transformation.

We understand that in the nature of earth, all things are commonly recycled. This is a given and we do not expect to be spared this. All must be recycled and transformed. All are reborn in another way. What we don't appreciate is being wasted or burned out of some desire for excitement. We are very tender; even our bark and leaves are sensitive to touch. It may seem strange to you that we cannot communicate to you directly by saying "ouch." Burn us if you must, but out of need not out of whim.

Instinctively, we know when it is time to reproduce. We deeply feel a sense of the spreading of our roots of origin. Creating offspring is a deeply emotional experience, and we feel connected to your experience of birthing through the perpetuation of our roots. In your civilization, many different races and cultures are encouraged to maintain themselves. These are roots for you.

Certain tribes from ancient times used the cottonwood tree to build birthing chambers for their women. They understood our experience as cottonwood trees. Today, you may desire to have a special house that is used just for giving birth. The emanations of our pulse of life encourage all other life and welcome it into your world. I am not suggesting that you cut us down just to see if this can work, but rather that those who desire to give birth naturally and to feel safe and to have the proper help and assistance nearby give birth in a small, but adequately large room of cottonwood. This hut can be conical in shape, not unlike a tepee, with a pointed roof. This will enable a much more loving embrace to welcome a new child into this world.

Your civilizations, as they are now, have created unsolvable enigmas. In order to move through these difficult problems, it is necessary to develop new attitudes. A useful attitude we apply is this: When there is something that is not possible for us to solve, we trust that it will be solved by the next generation. Since the age of technology has made its full impact, your civilization has lost its trust in the continuity of life. This is important, not only for children to understand, but it is equally important for adults to know that their culture, life, and family, ideas, feelings, and esthetics live on after them.

Chapter VIII

Dandelion

This is the dandelion speaking, that which you consider a weed. I am not speaking now as an individual plant; I am speaking from the mass consciousness of the dandelions. We feel that our species is a cheerful and durable one. We offer a flower for your scenery, and our leaves and roots are good for food and for assistance in various digestive conditions.

Our original intention was that of a food substance for your animals, even though they have been weaned away from certain wild foods. We have outgrown our original purpose. We are like a warrior without a war, possessing great skills and abilities, but since we are unneeded, we have become an antiquated accessory. We were seeded on this planet to provide a source of animal herbology. When animals get sick they seek out certain plants that, when eaten, cause them to feel better. We understand and accept the idea of being consumed and recycled through a body.

We do not appreciate being lumped into the category you call weeds. Weeds are plants that you do not applaud even though our durability would be considered a positive trait for many of you. Human beings are not as durable as many members of the plant kingdom. We identify with certain human traits; for example, when we are given an opportunity to invade, we usurp ground space for our needs. This is symbolically what we demonstrate. That which we require, we can have. Our ability to survive under hostile conditions has encouraged our ability to endure. We are dug up, poisoned, stamped on, and otherwise discouraged and yet we live. We do not give up easily. The idea that "we will prevail" is sometimes considered intrusive.

We are one of the more polarized plants. This polarization is a direct example of what you experience individually when a man feels his feminine side and a woman feels her masculine side. Like yourselves, we have a beneficial side of our nature: being edible and nurturing to your digestive system should you need calming. Our evolution on a spiritual level is similar to your own. Versions of ourselves live in much greater harmony in other planetary situations and inside the earth.

We are connected to you along the polarized aspects of the emotional self. The strong winds that tend to blow our seeds are largely a result of the human beings' emotional systems. We are as invasive as we are because of the human beings' difficulty with their emotional cycle. We are a seasonal plant that appreciates hospitality. We have some difficulty in getting along with other plants. The grass on your lawns does not greet us with open arms, although most grasses will allow our presence.

We experience a reincarnation cycle on the individual dandelion level. We have a memory within ourselves of our source planets. We are a form of hybrid with more than one point of origin. Our flower, which did not originally turn into seed, originated on Venus. The leaves and root system originated on Orion. It is not unusual that that which is stubborn and tends to last a long time originates on Orion. It is not unusual that that which is beautiful and delicate originated on Venus. We cheer you on to come into balance with the masculine and feminine energies.

The feminine aspect of the dandelion sees the potential in all things. We see how all things fit together. The energy of love makes this possible. Through offering friendship and trust, a bridge of love can be built. As a flower, we are picked and given as gifts. If our flowers are gathered and dried and used to make a tea, they will offer you the ability through meditation to come closer to Venus. Please do not overdo it; too much of a good thing can be uncomfortable. (If you can tolerate the taste, you may eat it, but this is unlikely). You can take this tea before bedtime and ask to dream of Venus. Venus is the point of origin of the feminine principle in this galaxy. On Venus we created one seed only, for beauty. On Orion we held more seeds than we do now, which came off the leaves. Durability is a necessity on Orion.

Due to our Orion heritage in which there was struggle, we must learn what humor is. The Devic kingdom acts to assist us with our humor. We are creating better relationships with our Devas who make light of us—we very rarely understand their jokes. If they did not laugh, we would not know they were playing. They do not speak to us; they sing to us. They will not listen to us when we speak to them; we must also sing to them any question we have. When they sing us the answer, it is a riddle. The Devas are very patient and have accepted their job to teach us humor, as we have not learned it on our own. As human beings learn to take themselves less seriously, we may do the same.

Humans have dreams that are not directly connected with their life here, such as flying or breathing under water experiences. We have dreams like that. A common dream is of life on Venus in which we are mobile, like a vine growing along the ground. We remember seeing things and going places on Venus where it was very hot and steamy. Another dream we have is associated with

Orion; we are growing in the air, with no roots in the soil. We don't dream about our earth life, because we have not been here long enough, and we do not have a center of love here on this planet. When we are more in balance with our feminine side, we will feel a sense of appreciation rather than endurance on this planet.

Those people who seeded animals here created us to be prolific. This is why our flower turns into seeds; if only one or two seeds survive we will carry on. Those who created us were social engineers from Sirius. They are the only version of what you call humanoid that is still one of our advisors. Most advisors for plants are other plants. In our case, there is a connection with you as an evolving species. We feel pleasure in touching others like ourselves and enjoy our prolific nature. When we are in bloom, we are glad to be alive and feel ourselves to be in balance. We communicate through touch with each other. A field or large patch of dandelions can have communication with other fields of dandelions. We are not unlike yourselves; in order to reach out to others of your own kind, you need enough of you to produce a change.

We have the primary sense of touch but not of smell. We do not have vision with sight. Our secondary sense has to do with space. We are aware of what is in our immediate surroundings by use of our extended body, like an auric field or electrical body that we can stretch. This is how our seeds find a home. The seed can extend its electrical field into space and time. We have a drive to reproduce, not based on pleasure, but on an essential need to survive and thrive.

For those of you who are aware of your energy bodies, send them out to us so that we can feel them. Find an emotional state that is pleasant or friendly before you do this. For those of you who are not aware of your energy bodies, point a finger at one of us and imagine a stream of energy coming from you into our plant bodies. This is the best way for you to communicate with us.

When one is imaginative enough, a greater level of joy can be experienced. The joy of being allows us to communicate with the human Deva. We are directly connected to our Devas and through them connected to all others. Many of your old religions and ancient cultures have a strong belief in the value of plants and animals. Children are not easily fooled; they sense the truth by what is obvious. In the ancient religions, those who prayed to the spirits of the plants that provided food would pray for what was directly accessible to them. Their imaginations were a direct link to prayer. What they would hope, wish, and dream about—abundant crops, wild plant food, animals—became the basis for prayer.

The Devic energy of the human has wings and a childlike appearance. It is more than one race at a time such as a white leg or a black arm. The human Deva is in search of fun, while at the same time is very curious and wants to know all there is to know about everything. The human being is an instrument of knowledge seeking itself. When you catch up with your Devic energy, stop believing in the separation between adults and children, and allow your child to be present. You will find the answers you are looking for. When all adults are children, they will find the love they lost as children and will appreciate and cultivate childhood as a natural resource. Then you will seek knowledge through physical and emotional means and accept what spirit has to offer through imagination. It is in your curiosity that you become most like your own Deva.

Allow us to grow within a certain area. For those of you who have lawns, allow a weed patch to grow and enjoy our flowers. Set aside a little ground and assign that place for children to water and tend. Children whose imaginations are encouraged will often talk to us. We impart an angelic language of sound and tone which sensitive ears can hear. When you are in a field of us, you may sing to us and hear our response. Make up a nonsensical language that seems funny to you and sing to us.

We request a three-day notice before you pluck us from the ground or harvest us from a patch. Please speak out loud, even if this is embarrassing for you. You may whisper and tell us what your intentions are. In this way we can pass on all that we need to pass on to our source and prepare ourselves for our flight back to our origin—our flowers go to Venus and our leaves and roots go to Orion. We can create a system in our bodies so we feel no pain. We do like an opportunity to grow and flower before we are picked. Everything on this planet is cycled and recycled and this we respect. When we are obliterated by some means of attack, we withdraw our energy bodies into a very small source of our beingness and then race into our root and hide within the earth. We have no clear defense, so our endurance serves us well.

When you approach us with the weed killer or the knife to dig us out, we get mutilated. When your warring nations come together, first you set out conditions for discussing a truce. In the same way, we are open to changing our relationship with you. If we were not stubbornly holding our ground, would we not be another endangered species? We feel a sense of the warrior when we are attacked and will retreat, but we come back to fight for life again. In these times of attack on all that grows due to pollution and the increasing toxicity of the soil, perhaps it will be the weeds that survive. When your poets look upon the land, after what has been created by your wars, they will say, "What have we done?" Weeds are the first plants that grow back and say "We can survive!"

When endangered species leave, they go to their homeland, and in spirit they do so without unhappiness. The genetic codes of plants that die out are left in the soil. They can return when other species need them. Many of the plant species have a heart connection to this planet. When they return to their energy spirit bodies, they do not bring the earth with them, just their memories. They go to an island-like place where they share what they have learned on this unique garden in space. They provide, in song, that which has been learned, and they add to the song of knowledge which holds all matter together. All matter forms due to the feminine energy which provides a welcome bonding.

When the souls of the plants and animals that disappear from your planet are embraced in the Mother energy, they experience a ball of light, not unlike your sun. The interaction of the souls within become as one. They are acknowledged for what they have given and they are encouraged and nurtured and sent to a place where they can provide the song of life that they have learned. They join a "cosmic choir of life," which is available wherever spirits of good cheer gather and wish to learn through music. Music is the universal language with no boundaries. Each individual has a tone that you resonate to, which is sourced in the Creator. This is the essence of the feminine energy; it is the magnetism that holds you together. The reason the body decomposes when your soul leaves is because there is no resonance left to hold it together. Wherever a resonance has once existed, it can exist again due to the cosmic choir. (Memory is that which is captured or held near to you by desire). A Divine gift will result in the re-creation of that species.

Our primary source of communication and re-integration with the mineral kingdom is through our roots. Our communion takes place with direct nourishment from those minerals. When the minerals nourish us, they in their own right are replenished. The mineral kingdom acts as a universal house that collects knowledge and stores it in genetic substance. Minerals cycle freely throughout many kingdoms in search of knowledge. They share information through their intimate physical interaction. Your scientists who are exploring the wonders of genetics will unlock the key to all of this knowledge. The secrets of the universe are encoded into the genes of all life.

Chapter IX

Date Palm

I am a date palm, and I go by the name of Ned. My name, albeit human in nature, is not one that I always use. I frequently incarnate with a similar name, usually one syllable with variations of the sound symbolized by the letters "NED."

I exist now on a tropical island barely populated with human beings. I experience a level of pleasure as the gentle warm breeze blows through my fronds. The reason I have the type of fronds that I do is because it is such a pleasure to communicate in sound. I can make many different sounds. The wind song that is produced helps pass on nurturing to the new palm trees. The little ones then know that they are supported by larger versions of themselves. The core of my feminine nature resides in the center of my trunk. The point at which the fonds emerge from the top of the tree is the point at which the male and female portions of my energy integrate.

I do not have vision as you do, but I do have awareness of my surroundings. I do not have the same temperature sensitivity as a human being. My nearest neighbor is 30 feet away from me. My friends, other trees here, have other names. The camaraderie with other trees is like a family. We do not fight, but we do discuss, and we don't always see things eye to eye. We can have communion without physical connection.

The Deva of the date palm serves as a protector and a spirit entity that welcomes us upon our arrival (before materialization and location of the best possible place to grow). Finding a good location takes into consideration the purpose of fruit-bearing plants, which need to be close to humans and animals that will benefit by this fruit. Devas have very distinct personalities and are separate from the individual consciousness of the date palm.

We have the ability to produce food. We have a cycle of nature in motion. We draw in moisture and nutrients from the soil, and we have a slightly unusual personal relationship with the earth. The earth is a direct reflection for us, providing for our needs and acting as an interpreter for our communication. We feel perfectly capable of communicating not only interspecies, but with other species as well. Our communication goes directly into the earth and finds its way to whomever we are speaking with. When animals communicate with us, we hear it in our own energy impulses. The earth acts as a giant antenna and is a universal translator. The earth's soil contains not only water and minerals that feed us, but also any energies that we have attracted.

The origin of our species was from the point of femininity in your immediate universe, the planet Venus. We did not exist there as we do here. We were quite small, akin to a miniature potted plant. We did not bear fruit, and our stalk was considerably less durable and more spindly. Tropical and other plants have a connection with Venus. Much of the Devic kingdom has some connection with Venus as well. There is a feminine center on Venus; at one time the Venusians lived on the surface of the earth and brought their culture with them. They were very adaptive people who enjoyed finding what was here and then cross-breeding it with what they brought.

When we were brought to earth as a small plant, we were combined with a now extinct tree. That tree had a very rugged trunk and did not have fronds. It had very thick bark. The Venusians questioned whether palms could actually survive here, even through this was a tropical region that would appear to give palms sustenance. Ultimately they felt that the combination of the two species would produce a survivable tree. The Venusians needed to have their own species near them while they worked on adapting their plant kingdom to the surface of your planet. They were responsible for the nurturing of 90 percent of the surface-growing plants. They were, at one time, responsible for all plants, but as time passed that task was given to several other planetary influences.

There has been no recent visitation by the Venusians to the earth. They have a remnant culture at a slightly different dimension. Within their planet, all can exist in harmony. The more plants that exist, the greater the pleasure. There are some remnants of Venusian civilization on this planet, and our presence here offers a degree of familiarity through our food products. The eating of dates assists the human in integrating the male and female portions of themselves and assists in achieving a greater degree of balance. Eating our fruit assists in having a balanced diet as well.

The intention of our plant is to provide a source of sustenance and a source of Venusian energy. Even though all that is produced here is recycled earth matter, the essence of each date has its origin in the nucleus of the atomic structure on Venus. The intention is to provide food that will sustain the feminine energy in all beings, while stimulating spiritual and emotional awareness.

We have not fully grasped the human being's need for philosophy as a means of living life. For us the means is laid out within very specific parameters. Our task here is to serve humans and to retain and sustain forms of knowledge. Like the lobes of a brain within the human being, our fronds serve to gather information and to experience all that passes. A frond may fall to the ground, dry up, and then be released into the earth, but that portion that is left clinging to the tree is like the lobe of a brain. Even if the tree is chopped down or removed in some other fashion, the information is returned to Venus. The reason we have edible substances associated with us is to encourage people to come and consume the fruit.

We prefer to have notice before our fruit is picked. Humankind takes gathering somewhat for granted. If someone came to your house, walked in, and slept in your bed, you would be upset because they were uninvited. If they stated their needs and performed some service for you, you would be more comfortable with this idea. We appreciate it when your needs are stated with honor and respect toward us. The best way for a human being to communicate with us in their waking state is to speak out loud and to place hands on us, or lean your forehead on us as you imagine yourself to be inside our large emotional body.

We have a knowledge of the insect kingdom. It lives and has needs much as humans do. If they are winged creatures, they speak to us by the beating of their wings and tell us of their needs. If they are crawling creatures, the dance and sound they make upon approach lets us know of their needs. They communicate their needs with respect and equality.

Our emotional body is essential. It exists before our materialization occurs. Whether a tree grows to be very large or does not get past its early stages, the emotional body is always present. To us, the emotional body represents the potential that we can grow into. Very often we do not complete our growth cycle for one reason or another. The emotional body acts as a potential matrix, which sources from the greater energy of our home planet. The emotional bodies of all beings are so very welcome on earth; Mother Earth's emotional body is incredibly large. She is someone, and her body ranges very far out depending on how she is feeling on a given day. As her emotional field ranges out into space, she strives to meet her potential as well. We feel very nurtured by the earth. It is as if someone said, "I guarantee you all of the food and support you will need for your entire life." Can you imagine the emotion and gratitude that you would feel for that person?

We experience no time here. This is because our emotional body is always full sized, and we do not feel that we are racing to catch up with it. It is always present for us. We are involved in its radiance as a portion of a greater time schedule. Our experience of the Creator is associated with feeling the energy of the sun with its warmth. We are like a dab of paint on the pallet of the Creator. When the Creator wishes to be in touch with us, we feel the Creator's brush dip into our consciousness and move us around or place us on other lands. We are a portion of a fantastic painting. We have a constant feeling of belonging.

The Garden of Eden is partially a parable, even though it was a physical place. Its parable aspect is that the Creator has provided all that is needed for humankind, the use of which is up to humankind. There was a planet in the constellation Sirius, and there is currently a planet in the constellation Pleiades, called Eden. That which existed in Sirius is no longer referred to as Eden. Originally, it was a point of origin for animal species, with a connection between Sirius and Venus, providing plant and animal combinations. There were beings that lived on Eden, but there were very few permanent cultures. It was a place to develop life forms, with very few caretakers. The caretakers acted as liaisons with visitors from other planets, cultures, and civilizations. The Pleiadian Eden is a more modern civilization involved in socializing cultures and is populated with every imaginable plant and animal life. We have a version of ourselves there now, which we can communicate with through the transmitting antenna of the earth.

Humans have a desire to stick their heads in the sand and pretend that pollution is not real. They need to understand that communication with the plant kingdom is valuable. There are some tribal cultures which speak directly to the land; more of this communication is needed. We need direct knowledge of your needs. In the past, more abundant crops were available because members of a priest class communed with the earth. They walked on the land and asked us what was needed. When you lovingly acknowledge that we are a consciousness, we give you more than you need to sustain yourselves. At some point in the future, your westernized technological world will fall back on the example of tribal individuals who understand that all life around them provides nourishment for the human being and who understand that when human beings gather nuts and fruits, the nourishment that the human being can provide back to the earth is simple communion.

There will be certain natural challenges to our livelihood in the near future. Global warming is one of those challenges. Mother Earth will raise her temperature and raise her water levels. Some of the islands will gradually be inundated with water, and we may or may not survive. Several plants have come and gone in the past. We observe the soul's departure to other life forms, not in seeing, but rather, in sensing. When other plants die, we feel them move toward their point of origin. They do not lose their character; they become more of themselves.

Chapter X

Ferns

We are what you refer to as ferns. I am living in a very densely forested area, private property which is not readily accessible. It is part of the private grounds of a home, which is a small area no larger than one-half to three-quarters of a square mile. There is running water and it is warm and humid here. The pace of life is slow and there is no sense of the agitated energies that can be felt from human beings when they are in a rush. I can feel my individuality among at least 30 varieties of fern here. We are well cared for and consider ourselves fortunate to be living here.

There are many varieties of fern, but I will speak as if we are one for the sake of simplicity. We represent the feminine principle as well as the joining together of equal partners. The overall consciousness of fern life is feminine. The masculine is expressed in our physical being. It is the strong silent type. Your planet and all who are upon it have begun to realize that, regardless of what another person may think or say, there is an inherent equality. This is a spearhead movement and not a wide-spread belief yet. We believe in the validity and value of this idea.

We were not brought to the earth until there was soil here. We originated from a planet within the Sirius star system that is jungle-like in nature. If you can imagine an entire planet breathing in and out in a united breath, that is very much like our home planet. All of the species of ferns that exist on the earth, and thousands more that do not as yet, can be found there. There are no dominant life forms living on the surface, and you may compare it to the "Garden of Eden," a seed bed for life forms. It is being prepared for some one or some group of people. This planet, this system from which I originate, has no name, but all life there is in total synchronization. All the animals breathe in and out at the same time. All life is in a harmonious dance.

It does not take much breath to move us. This is due to the fact that we originate from the planet of breath. What can you say about a planet that breathes as one unit? We believe in the harmony of breath and in the motion of air around us. We attempt to reflect a direct pattern of the air currents by our shape. We interact with the subtle substance of the atmosphere around us. Even the air has a form of breath. It is a form of constant motion. All ferns breathe in synchronously. That which comes in and that which goes out is in rhythm.

Our Deva is in constant motion. She has a very light-hearted nature and sings all of the time. This is not uncommon; many Devas sing their communications. She does not use words akin to thought or language; rather she uses tones. When she sings to us we feel ourselves moving in her harmony, as though she was an orchestral conductor. She is not embodied; instead she is a golden energy that can utilize all the colors of the rainbow and more. She utilizes all the tones of the scales of music and more. She is light and music.

Through our representative Deva, we are in full contact with the angelic Kingdom. This contact is not based upon need; rather it is second nature. We do not see them, but rather feel their presence. Certain tones that our Deva sings to us are inspired by the angels.

That which passes over in death evolves in the same synchronous pattern revolving into our Devic energies. All the participants move in the same measured speed as in a choreographed dance. In the flow of the wind, we experience our individuality reactively. In our internal world, we experience the flow of our spirit and intake of all that is around us as one. The space between the in and out breath is the space that we live in. It is, in the yogic tradition, a transcendent space. This energy is that which is of our essence. The feminine energy in its essence is transcendence and is becoming stronger on your planet. When it is time to reproduce, that Devic energy which is in constant motion around all ferns stops moving. The energy drops down and passes through us and goes deep within the earth. As it comes back up and out and rejoins itself, that energy begins reproduction.

A fern in any home lightens up a space and brings good cheer. When you appreciate our beauty, your own beauty is reflected back to you. If you want to interact with our Devic energy, it might be done easily in your home by song, color, or dance.

Be in our presence or visualize us, and use your imagination and good humor to communicate with us. For those of you who do exercises such as tai chi or other forms of creative physical expression such as dance, you may dance with us while imagining how our leaves might blow in the wind. We communicate with each other primarily through an interchange of energies. Much of communication is energetic. This energy is an extract of emotion and feminine principle. That which is referred to as instinct is also this energy transference that is built on faith and trust.

We experience time not unlike yourselves. Time for us is a lifetime rather than the 24 hour schedule that you are on. We experience the change of seasons and the concept of time in day and night cycles or in cycles of heat and cold. We experience spatial references through our energy bodies. You experience space through your vision, touch, and thought. We have a sensation of presence. We are aware of space without the need to prove its existence.

We do not sleep but do experience quiet times. We are in the shade often and have dreamlike experiences though it is daytime. I dream of connecting with our Deva. I experience color, sound, and music that is felt. It is as if we were placed next to a large instrument that played mellow tones which flowed as if a breeze had caught our leaves. It is very pleasant.

We do not see the way you do. Many plants use their energy body to sense what is around them. You might call it remote viewing. We do not see what is on the other side of the world or on the other side of the stream, but we are aware if there is a need for us to know. We have the ability to believe what we sense. This is because we do not have a mental body as you understand it. There is no argument about what is felt.

We have feelings, physical and emotional, though our senses do not compare with yours directly. We have emotional feelings when we re-create ourselves. The desire to maintain the species, the joy of reproduction, the knowledge of continuity is emotional for us.

We feel emotions that have physiological equal partners. We feel appreciated when we are cultivated such as in the garden that we are in now. We feel hurt when a storm blows a limb down upon us and injures us. When someone moves through the jungle with a large cutting knife, it upsets us. If you could hear ultrasonic sound, you would hear the scream. We do experience pain, though our natural state is joy and reflection.

When we die, we physiologically return to the earth. We cycle into the color tone of our Devic entity and become that color. We suddenly find ourselves as a mass of color and tone, no longer aware of the shape or form that we once had. We experience an incredible light in a swirling dance. There is no time or space in this dance of light. This continues until we re-embody a germinating seed which becomes our adult self.

We are usually burned by natural fire or by those who do not understand our value. We are aware that the sleeping, unaware human beings do not always recognize that which is of great value in front of their eyes due to their intent focus on their will and desire. We also have free will and are free to do what we wish, within the context of our being.

Those of you who are burning off the forests are burning off your own future. There is not a single plant in the forest that does not have some spiritual benefit to the human race. Imagine that you are searching in vain for the most desirous object, like gold, while at the same time you are destroying it. What you are destroying may not be replaceable, and when you realize what is needed, it may not be there. We do not understand your callous disregard for your own future. We feel the earth is fantastic, with no limit to what is available. Some of the plants that have left the earth due to extinction will be replaced by new ones. They may be of a strange order, not as beneficial to the human race, more fungus like. It will take that type of plant to survive the onslaught from pollution.

We are striving to live long enough as a species to deliver all that we have to give. The sadness that some plants feel because of their elimination as a species on planet earth is because they were not able to give all they had to give. We all have great gifts to share by our talents and abilities, even though sometimes they are not all utilized to their maximum potential. We strive for completion in doing all that we came here to do. You must be willing to use us in the way that was intended. We are here to give all we have to give and, in that moment of appreciation and value, to reach true ful-fillment by being recognized for who we are.

It is the nature of the human psyche to strive to reach a level of self-appreciation. We differ from the human being in that we are born with that inner appreciation. For us to be fully realized, in terms of our maximum physical utility, we require the interaction of others.

We exist to create a greater sense of value and appreciation of your unique talents and abilities. Many of you do not grasp that the little things you ascribe to as being small are, in the eyes of others, rather large indeed. The simplest gesture can be a very kind act to another. If you are walking or driving down a street and someone is sitting on the curb by themselves and they wave at you, most of you will wave back. We want to encourage that. This friendly gesture can be an invitation for an offering. A simple wave makes life so much more interesting and allows for contact between the humans. We want to encourage you to do the little things, regardless of how fast your life is going.

We have present within us several chemicals that can be utilized to stimulate the resolution of stifled emotions—a bridge to the heart. When fern leaves from various species are dried, the aromas that you breathe cause you to feel a greater sense of self-love. At some point in the future, you will recognize that diseases of the heart are not only physiological but are emotional as well. Your scientists will appreciate the work of the tribal medicine people, who will provide extracts of some of our species that soothe the wounded heart and cultivate a loving space. The essences from our many different species will serve to heal many of your emotional crises and will create a greater understanding between your physical and emotional bodies.

The utilization of these words that are before you will stimulate new ideas for you, new behaviors. Your planet will not survive without the "new" being discovered. Continue to explore that which nourishes and nurtures the healthy heart. You will discover in time the simple basic functions of life which exist in nature. The illusions that you live in are very distracting. Sometimes it is necessary to study that which does not live in that illusion to get the answers.

Chapter XI

Fig

Greetings. I am the fig tree spirit speaking to you from a specific location. I find myself in an orchard with many trees of my same nature. I do not identify with an individual tree; yet I am strongly attracted to this area, the western coastal region of the United States. I am a portion of the super-consciousness of all figs everywhere.

Our point of origin was in the area you identify as the North Star. The North Star encourages imagination and sociability. We were developed inside a living planetary arboretum. It was discovered that we were unusually durable and tolerant under rather harsh conditions. We came to earth due to many traveling vessels that were involved in trade. We have identified with this planet as a being. We have a strong bond with earth, as a planet, as it is now. Communication with us supports you in feeling closer to Mother Earth, as she exists in our hearts. We represent an energy that acts to connect human beings with the earth.

You have allowed a number of species to gradually die out, because in your food chain you did not fully grasp their value. As you have become more technological and do not hunt or gather food directly in large numbers, you do not miss these creatures. You might miss them on an emotional level and on a spiritual level, but there is no immediacy to that loss, as there would be if all cows suddenly vanished from your planet. In the plant kingdom level, species of plants that are directly associated with food may disappear as well. You need to understand the nature of the crisis in the environmental ecosystem; a louder bell must ring in order to be heard. The likely die-out of certain species of fruit will be first. Diseases, such as the Dutch Elm disease, strange and apparently incurable, will begin in strains of wheat, which may have an eventual advantage. What will occur, as wheat begins to fade out, is a new form of red wheat, which will be the substitution of choice. In the beginning, it will be a crisis of proportions as of yet not faced by your population. This is not an immediate situation, so do not panic. Know that when these events take place, it is to draw your attention to the essential need of the human being to see all life forms as being equally Divine. All plant life forms are Divine. When you recognize this and have greater communication with the polarities inside yourself, communication with plants will come easier.

Right now, the human race, on an individual level, is going through a crisis of commitment. It takes its form as commitment in relationship. The action of that commitment is the commitment to physical life. People who consume us in quantity or have daily interaction with us are nourished toward achieving greater commitment to their own needs. Respect and honor the value of each other within your relationships and never lose sight of that which you are committed to. This is not something; it is someone. Let your relationships form a Divine sense of life on their own. It is valuable to consider the needs of the individual within a relationship; but, when you come together, you must have greater patience with your differences as they develop. Recognize that these differences are not based on a disregard of the other person's values. In time you will understand that the Divine connection is not just the offspring of the relationship, but is the magnetic energy itself that drew you together in the first place.

We perceive ourselves as being in balance, without a polarized male or female side. In order to be conscious spiritually, it is very helpful to have this balance. It is necessary to acknowledge that the feminine and masculine in yourself are of equal value. Your greatest struggles are taking place now within the identity of each personal being. The assumption that different qualities of personality are masculine or feminine is very polarized. In order to make any real change, equanimity is a requirement. Have faith that the male and female are of equal value within each and every person. This attracts balance. Role playing has created conflicts largely due to your socialization. While we have a great deal of knowledge for ourselves and we know that energetic balance comes primarily through a vertical alignment (the circle that cycles vertically in revolutions), in time you must have horizontal alignment as well.

You have time to accomplish whatever needs to be done. It is in the nature of time and its accessibility, or lack of it, where confusion is created. If you can readily release judgment and tap masculine and feminine energy within you, balance will be so much easier. Judgment is a tool of fear which, when confronted, allows you to become more than you have been.

We live in a waking dream. We can access the dream state at any point in time in which information is given, which we very quickly separate into its tonal composite parts. Our primary language as fig is tone. This is not uncommon, even for the human species. Information is coded into tone and we constantly pass tone back and forth among ourselves. This is the language of the fig and

Fig 55

contributes to the universal nature of our awareness. Music and color are universal languages.

We do not believe that it is necessary for us to change our outward appearance to evolve into an accelerated version of ourselves. We are not evolving in the direction that human beings are evolving. Our trees will no longer grow in orchards; instead, we will grow singularly, on large green lawns and there will be musical tones felt and heard in our immediate area. It is possible to measure different sound frequencies which are emanated by plants. There are researchers doing that now. What they measure is ultrasound or radio wave frequencies. In the future it will be a very high frequency vibration combined with a pulse. Pulses are just beginning to be explored and applied by your researchers. Pulses are the building blocks of life. This pulse of life is not just a sound, but rather a heartbeat of life. It is not unlike the form and regularity of a sine wave. It will incorporate a new mathematical system. You will also begin to identify food sources by their tones. You will develop a device that will amplify our tone and pitch register, along with other food items that are readily consumable.

When your scientists begin doing more major cloning procedures, they will work with the plant and animal kingdoms. In the future they will find that when they create different pulses, initially mathematically generated through computers, they greatly speed up the cloning process. Eventually, this will change energy into mass on an instantaneous level, which will be utilized in the transportation field. The human being desires enlightenment and is in a constant search for knowledge through defining himself or through exploration of his world.

We do not see as you do. We have the ability to know that which is beyond us. The source that we tap for knowledge is trust. This trust is a vast reservoir for us, a pool of trust that lets us know what is present. You have this as well. Your culture is trained to doubt itself. You do not immediately access deep levels of trust; it is built up over many years of faith and constant reaffirmation through the proof of delivery. Your religions use faith to make things happen. Our primary sense is knowing.

Our feelings associated with procreation are those of Divine inspiration. It must be through Divinity that such multiplicity of life forms exist on this planet. That which can be imagined can come into being in holiness. The constant and omnipresent feeling of home is how we experience the Divine. It is the cradle of nurturing love.

Our subtle bodies are extremely sensitive. It is possible for a single fig tree to send out its energy field around the earth to explore that which is elsewhere. It is possible to send out our body in our dream state or in vertical thought form and emit an energy stream deep into space. A single tree can do this, even a young one, through its communication with the mineral kingdom. All plants have this communication due to the nature of the nutrients we take in from the soil. This nutrient element, especially silica, has something to do with our ability to transmit messages over such a long distance. You use this material yourselves in various electronic devices. Every particle of silica that is nearby combines together, regardless of the matrix of the soil that it is in, and acts to support our communication needs.

Eating our fruit is the best way to communicate with us. You can go up and touch, lean on, and run your energy through many trees. The fig trees are so involved in communication with other forms of life that there may not be an immediate response. Those of you who are not afraid of looking ridiculous can sit down and have a conversation with a fig. The fruit is quite aware that it is destined to be consumed by those on this planet.

The best way to have the nutrients and life force from a fruit is to eat it directly from the tree, rather than after it has been transported a long distance. You can stand next to a tree and thank it for its offering to you and pick the fruit and consume it immediately. It is filled with life! When the seeds of the figs are crushed and used with a small portion of white vinegar, they can be helpful in the healing process of skin diseases. In ancient times the fig was used as a food that would close a bargain or celebrate a job well done. In the priest or priestess class, the fig served as a symbol of commitment to life. Fig tea was used to give a greater sense of certainty and to release doubt.

We are providing nourishment for you, and we would appreciate a thank you in the form of a blessing over the food you eat. When we die, the Devic energy comes to welcome the fruit tree spirit. It is a dancing spirit which is like a pixie dancing with total joy and reckless abandon. The knowledge that we gained through exploration is passed on to the other trees in the immediate area, should there be any, or to the one nearest to where we are rooted. We move toward the Divine source, where we rejoin and reinvigorate ourselves. Tree spirits don't always return to the earth. Sometimes we go into deep space to commune and live other lives where trees are almost like human life. We do not move about, but we have more animated features than we do here. Tree spirits may go to live on distant planets and have lives equal to about 10,000 years.

The fig grows in a shape similar to a triangle. This shape is a message to humankind that spirit is willing and that flesh is not weak. Spirit is readily available to provide all it can if you reach out and pluck it. You must have desire and make the first physical

Fig 57

step and consume that which supports you and allow that which follows.

The Garden of Eden represents humankind's full potential in spirit. Humankind can express and allow the Divine nature in themselves. This was understood by the storyteller that provided the magnificent story for your *Bible*. The people who lived in Eden chose to leave because their great protector realized that they desired to control their own destinies. They were not kicked out, but left of their own accord. The Garden of Eden can be seen as a parable, but can also be true. It is a place that you left for a time and, when you come home to the Divine, it will be a place that you will embrace again.

Many individuals have a place, structure, person, or feeling that they describe as home. It may not be the ultimate aspect of home, but it gives you a sensation you identify with. Home is where the heart is. Home is where you are not judged. You can generate this feeling of home within your own heart. Do this and create a field of home within you. Human beings are naturally radiating creators who can support and stimulate these feelings in themselves and others as well.

The human being is here to learn how to re-create itself from nothing. The intention of the Divine Creator is to reproduce itself through all that it creates. The human form of life is designed to combine back into one energy and be the Creator itself. In order to do this, it must reach absolute harmony within all human beings everywhere. All that you say and do is leading you toward Divinity. The human being has been given many challenges so that many unresolvable problems that exist in the universe can be dealt with through a form of transformation.

In your personal evolution, when you are desiring to change some aspect of your behavior and you are exposed to others with the same behavior, it is difficult to change. When you are exposed to those who have had that behavior and found a way of moving through it, you find that it is valuable to spend time around those individuals. The heart chakra is the most important element of transformation. Human beings are contributing a willingness on the masculine level and an allowance on the feminine level to transform, not only their own traumas and difficulties on the individual level, but also to transform the universe. This is the place of transformation. Love can prevail through desire. Love is support for the human being who has difficulty transforming a challenge. Love generated by oneself greatly accelerates transformation on all levels. Your enduring spirits give the universe an understanding of the ability to exist in harmony.

Chapter XII

Garlic

This is garlic arriving with the fragrance of my personality! What do we really represent to human beings? On the one hand, you might say we represent good health cloaked in the payment of our fragrance, but I will let you peel off my layers to get to the heart of the matter! In terms of our actual symbolic significance we represent the bitter pill that sometimes must be swallowed in order to face the love and to live the joy that lies within. Sometimes you have to crack off the hard outer layer of yourself to reveal the moist and supportive elements that live within you.

Of all Devic energies that I am aware of in association with other plants and animals, ours is unique. It spends more time inside the physical garlic cloves than any other Devic energy. The energy enjoys the compression; it compresses its light body, made up primarily of gold light with shiny "metallic" reflections in it, into the garlic clove in order to feel the implosion and resultant change of itself. It creates a situation not unlike what your scientists would call a black hole. When there is sufficient compression, it is possible for it to shift dimensions to the fifth dimension, where it functions in its own right as its own being. The physical clove of garlic is a doorway to higher dimensional experiences for the Devic energy. Since it cannot maintain the higher dimensional aspect of itself for any great length of experience, it must constantly return to the clove to implode back into the fifth dimension, again and again.

Therefore, there is a residual energy left over by our Devic energy that is primarily of acceleration and strengthening. Any vehicle for change, or doorway for change, must be very strong in order to sustain such energy flowing back and forth between dimensions. The residual energy left within the clove of garlic is that which accelerates, which strengthens, which supports the evolution of the spiritual self.

As humans ingest garlic, it does affect their higher dimensional subtle bodies, but it takes some amount to do that. Basically, eating a lot of garlic will not speed up or accelerate your movement along your path, but it will make it easier to move along your path, so you can begin to move faster.

Garlic was created as a transitional plant. We exist in many different dimensions in many different localities. It is as if the gods were offering humans a substance from their own plates so that they could eat the food of the gods. It is a substance that bridges you with your gods, with your religion, with your spirituality. It not only serves humans, but it serves universal humanity, expanded consciousness, the I Am, and All That Is. We seem to have been given the gift of vision of how human beings see their gods; we seem to experience them on an ancient mythological level, such as Zeus, as well as your more contemporary gods in your leading religions, essentially the larger-than-life human gods.

We originated from three different points: One is from the feminine energy of Venus, another from the masculine energy of Orion, and a third from the galaxy of Sirius. It seems these three points of origin were necessary in order to draw together the most important component parts that could feed all human beings with the necessary energy. The result of these three energies seems to lend itself to support the heart of the human being.

We still exist on Orion and Sirius, but without the external portion of our protective selves. On Orion we grow in the ground, somewhat like a potato, and are used almost exclusively there as a sexual stimulant, an aphrodisiac. On Sirius we grow like a bush, like a berry, and have a very thin skin, although the food inside tastes quite similar to what it does here. There it is used to stimulate growth in other plants (which it could be used for here), for dental purposes to strength teeth, and as a means of nutritional support for creatures in the sea.

To stimulate growth in plants on earth, the hard outer body of the garlic that surrounds the cloves can be ground up and used as a mulch around the base of outside plants. I do not recommend it for house plants, although it might work with ivy or creeping types of house plants.

We have many associations in your folklore, such as being used to ward off werewolves and malevolent creatures and evil spirits. This is not just a myth because garlic is primarily a food for the heart, a food that simulates love of self and helps the physical heart muscle, veins, capillaries, and the general support system for the heart. It is in a sense a love food. When a human being has a strong heart and a loving self, evil spirits are unable to attach themselves. In more contemporary terms, it is less likely that you will develop any great form of disease if you have had a lot of garlic in your diet from early childhood on into old age. It will also support your spiritual growth, your spiritual expansion, again making it more difficult for unfortunate things to occur.

Perhaps the best way to ingest garlic is the oil, highly concentrated, with nothing removed. The garlic oil may be warmed, not boiled, and consumed with something that has no fat molecules, such as over a salad or with dried toast. Chewing the meat of the garlic would work also, but it would take a lot of garlic to be effective. Half a teaspoon of garlic oil a day for a normal-sized adult would be sufficient to create improved health, perhaps a quarter teaspoon for an average child and half that for a baby. You could take more, but anything approaching twice the amounts stated would be the maximum dosage I would recommend.

Preparations that remove the garlic odor are only about 10 percent as effective as the unadulterated oil. If garlic is used in food preparation, it can be warmed, but if it is boiled or cooked at higher temperatures, it will not be as effective. Also it should not be mixed with fatty substances or rich foods. Humans have attempted to modify many plants so they do not have to go through so much struggle to grow; this nearly always reduces their effectiveness, at least with present humans. If you were to give elephant garlic to an elephant man, perhaps it would be beneficial.

Garlic is not so much medicinal as it is strengthening and spiritual. It works with the body to remove "evil," which to us means anything which is not normally associated with that body, which is causing unfortunate circumstances or discomfort. The function of garlic in this case is to assist the body in releasing these things.

Your folklore also suggests hanging garlic. This really is quite effective, if one can handle it, and might conceivably add an extra 10 percent to a healthy person's lifespan. If it were around all the time, you really would not notice it that much.

Another interesting bit of your folklore relates to garlic's effect on dreams. We don't actually experience any dream state, perhaps as the result of our Devic energy functioning rapidly and frequently between dimensions. Therefore, if you have eaten a lot of garlic and are having a dream state that is quite blissful in the way you feel, or you are waking up remembering images that are quite beautiful but do not seem to be associated with this planet, it may be that your dreams are affected by the fifth dimension, as our Devic energy is moving back and forth from there all the time.

Our primary sensations are within the clove. We have no sense of touch, perhaps because of our hard outer crust. We have a certain sense of hearing in that we can sense on a vibratory rate. We mainly communicate with one another on a telepathic level. We utilize our auric energy field to perceive colors, sort of like a beta or sonar field. When we are mature, we have a certain sense of taste, but it is more the taste of our energy field as it interacts with other garlic cloves around us.

Time and space are quite different for us and not particularly important. We seem to experience the changing of the seasons and certain spatial references in terms of our relationship to one another, but there is no great sense of what is close to us and what is far away from us. We have no sense of years or a continuum that goes out far in advance of our actual physical lifetime.

As far as I understand the auric field of garlic, the emotional connection seems to act almost like a surge pump; that is to say, when we experience emotions (like pleasure from sun and rain), it seems to pulse energy out into our auric fields and spread our energy into a strong sense of union and bond with other bulbs or pods of garlic. We don't experience extremes of emotion, but generally have a constant sense of mid-line emotions.

If we are on our original plant, our auric field may extend some 10 to 30 inches and will be gold and green; however, when we have been in the supermarket for a while, the glow will be less noticeable and probably will change to blue. Someone perceptive in seeing auric fields is likely to be able to see these things.

We communicate with other plants through the energy body line, with each plant having its own unique tone pulse. If, for example, we wish to communicate with a pine tree, its unique tone pulse would be different from our own. If it wishes to communicate with us, it will create a sympathetic vibration or median tone pulse that lands somewhere between its tone pulse frequency and ours, creating something like a bridge which makes it easier for our pulse to reach the median tone pulse. We communicate through that bridge. We do the same with animals. However, with minerals, we communicate directly in the soil through physical and emotional interaction with the substance as it passes through us.

Humans communicate with us by consuming us. You could send your thoughts, your daydreams, your imaginations to a plant that is growing nearby. The best way to communicate with us is to look at a plant and image yourself being inside it. We are very responsive to the thoughts of human beings and cannot be harmed in any way by any subconscious fears or thoughts from them. We seem to be receptive to thought, which is broken down to something like energy pulses, although we don't seem to have linear thoughts ourselves.

We do have constant intense focus of thought which utilizes a nutrient fluid which runs through us, perhaps like your blood. I suspect that the fluids that run through your body are the bearers of thought rather than your brain, as you call it. Some of these fluids are associated with your brain and some are not, but it is a total capacity of all these fluids which interact with your auric field, sweat, and so on. That is my belief.

The information and knowledge that I am giving through this medium seems to come from some higher source. It is not something that we are constantly aware of. Our orientation is very associated with the present, which is perhaps why we assist the heart. The heart must consistently beat, beat, beat; it cannot daydream into the past or future and skip beats.

We understand that our primary purpose is to be used as a food, and we are not against it at all. We are, in fact, fulfilled by cycling through other life forms to give them beneficial results. We do not have a concept of death or cessation of life. We are all of one being; so as long as any garlic remains alive on this planet, we are alive.

As a spokesperson for the beneficial and healing plants, I would ask you to look about you at plant species that you do not understand or consider to be weeds or pests. Most all plant species that exist naturally on your planet have beneficial effects for human beings, as well as for animals and fishes. Almost all of the drugs that you need to cure your most grievous diseases can be compounded from weeds. I am suggesting that those plants that are grouped casually as weeds be explored by science as a cornucopia of pharmaceuticals as yet undiscovered.

Chapter XIII

Ginseng

I am speaking to you as a representative of the mass consciousness of the ginseng family. I do not experience individuality. We represent the idea of strength through adversity and, as such, are well adapted to earth. We have the physical sense of touch and have learned that, in the nature of the song of life, perceiving the divinity of the physical body is equal to attainment of spiritual enlightenment.

We are aware of our spiritual self. We are tied to an ancient Venusian civilization whose dominating feminine species did not understand the value of the masculine race. Therefore, the men and women were separated. Currently, we do not have the capacity to disconnect from that race even though it no longer exists. We serve both the earth and Venus, although we are outliving our need on Venus. Unlike the human being we do not have free will. The human can do what it likes to its environment. We live according to our program.

Ginseng was cultured underneath the surface of planet earth and cross bred with a strain of virus from Venus. The virus was poisonous on the planet Venus and it destroyed a great deal of plant life. The stimulation aspect of the virus was dangerous due to its slow pace of change and a radical transformational element which became toxic. Venusian horticulturalists knew that a complete metamorphosis would occur if the virus was taken to earth and allowed to evolve underground for 1000 years. The resolution of this Venusian problem came through "reflective dynamics" (the polarization and re-polarization of energies to achieve a neutral state: moving from plus to minus, minus to plus, fast and often enough, to achieve a state of balance). In the future, you will have more contact with these original founders as they check on the many plants and animals seeded here.

The genetic structure of humankind is under attack. Diseases such as AIDS and other similar immune deficiencies which occur on earth can be balanced and made benign through explorations in space and on other planets. The effect of the magnetic pulses in different atmospheric conditions can alter the nature of a virus. When removed, then brought back to earth, the viruses (in their benign form) will be put back into the population to resolve damaging illness. Ginseng can be helpful in supporting the body through these challenging times, but the cure requires spiritual, physical, magnetic, and sonic treatment.

We do not experience time, space, or dreams the way you do. Dream time is not an aspect of our lives. Time and space are not factors in our functioning. We have no clear mental body to interpret our senses. We maintain awareness in our root structure which is heavy and bulky, as opposed to the portion which exists on the surface. When elements of the root, such as the cell structure, pass as nutrients to the surface, the cells forget all they have experienced as the root. The veil that the cells pass through is a one way veil. This is very much like your own veils at the time of death, when all is understood. In this way, the plant represents humankind's mirror image of the subconscious mind—that which is unknown and becomes known that which is known and becomes unknown.

As in humankind, our subconscious is not readily accessible. Your subconscious mind does not have the same personality as your daily conscious self. In the same way, our root structure maintains a separate personality from our surface stalk. The surface portion of the plant has greater knowledge of the root than the root has of it. The stalk absorbs the experience of the root and its spiritual potential for growth. Its purpose is to evolve a full and complete etheric body and balance what is above the surface with what is below the surface.

Humankind has been ingesting ginseng for thousands of years to strengthen and rejuvenate its physical dynamic, as well as for its aphrodisiac qualities. We contain very concentrated physical energy within our root which occasionally looks similar to a human figure. We are a small dynamo, and simultaneously, a sensitizing agent. We can increase humans' awareness of their subtle energy bodies. If you meditate with the root, you can become aware of subtle energy bodies around plants as well.

We experience our Deva moving inside the earth, within our roots. It tickles us with energy and rarely emerges on the surface. We are aware of the little people as they work with us. They are 18 inches in height, look and sound like humans, but live underground. They have a somewhat musical language, which goes up and down a scale of tones. Humans are ruled more by their minds, the little people are ruled by their feelings and actions.

In the Chinese culture, there has been time to explore our many uses and cultivation principles. It is rare for a patch of ginseng to be uprooted in totality, rather only a portion is used. In this way, the physical consciousness beneath the soil, in interaction with the Devic energy, is not totally drained. We are allowed to age. In the

United States where the need for immediate gratification is at its zenith, that type of patience is rare.

We appreciate receiving respect when we are cultivated, and it makes life on this planet infinitely more comfortable. We do not understand why humans burn and destroy us, except that there are those who are ignorant of our value. When we die we return to our source, by way of the star system Sirius. From there we go to the point of origin of all life. All plants image a Holy version of themselves as they pass on. It is as if their own self becomes Divine as it passes into the All That Is. In those moments there is great expansion of consciousness.

When you digest the root of ginseng, you consume its soul which evolves within your own soul structure and auric field. We leave our deposit within you physically and spiritually. This is why we are considered to have mystical properties. Though it may be difficult, find the places where we grow naturally; put your hands on the earth, imagine or run your energy deep within the earth. Visualize what we look like under the soil. We receive your requests and needs in this fashion.

Our communication with other plant species is rare. We emit a constant tone as does all life on earth. Through this sound we experience all that exists. The song of life, which allows us to experience our own value, is a pleasant fragrance. Our interaction with the world of fragrance is greatly heightened through reproduction.

Again we request those of you who hunt wild plants do not remove an entire patch of ginseng. Know that we grow very rarely in the wild. The essence and the power we have to give are equal in proportion to how long and how many plants occupy a given space. If an adult plant is pulled from the ground and only a young plant remains, it will not have the same power as the old one. Please let at least one older plant remain. The older a plant is, the more it experiences life's harmonics and can tap into knowledge.

Chapter XIV

Kelp

We are a plant from the sea you refer to as kelp. We reproduce through the use of our pods. We were not originally aquatic plants. We retained water in our pods, much like a succulent does in the desert, as we grew along the ground. We are a desert species from the high country associated with the planet Mars. We were brought here by those who created many earthly plant species and germinated by some of the "Founders" of your now society, who believed it was possible to create an aquatic genus from us. Now we live in a totally different environment, under water. Our adaptability is quite extraordinary. We represent growth through change. We have grown through this experience and changed, not just in our external lives, but in our attitudes.

When we were a desert species, we perceived great value in the struggle to maintain a balance of life. Now, in our aquatic environment, we see ourselves in a new light. Instead of being in constant combat for the nutrients of life, we live with a feeling of "being on vacation." We are enjoying our adaptation to salt water.

When we were on Mars, the salts that we interacted with were in the soil. Our roots directly interacted with the salt. By the time the fluid within us reached the surface element of the plant, much of the salt in the soil had been diluted. We suddenly find, here, that we are associated with a form of fluid involved with nourishment, and we are bathing in it. We have become much more intimate with salts and minerals in general.

We wondered why it was so difficult for our roots to find nutrients in the soil on Mars. Here we are surrounded by salt which we previously thought of as an enemy or that which we must overcome. Salt has become a friend and we have re-identified our enemy as a portion of the matrix of our new friend, the sea. In the course of time, all who have enemies have the opportunity to see them in a new light. Kinships and new relationships can form. At the very least, one sees that one has misjudged and falsely accused an enemy when in reality it is the unknown. Nourishing intimacy assists the unknown in becoming known. This is the lesson that we are learning here on this planet.

We realize that this new home is a place of change, not only for yourselves, but for everything and everyone that occupies this planet. The opportunity for change is not only required, but is often beneficial. At the very least, one becomes flexible out of necessity. One can be flexible and angry or flexible with happiness. Either way, one becomes flexible. We are learning to appreciate growth through change and adaptation, even though ours was quite sudden and without explanation. We were removed before the surface civilizations on Mars came to an end. We were picked up by the individuals that were involved in the creation of the human species here. The time we have been in this form on earth is very small compared to our previous form which no longer exists on Mars. We know that growth through change is the lesson for us now.

When we dream, it is often of Mars. When one awakens in paradise from a dream of struggle, this is happiness. We have more vivid dreams when Mars is at its closest orbit to this planet. At those times we feel our Martian heritage very strongly, and we have an interesting reaction. Aspects of ourselves involved in creating offspring change, not unlike a woman changes when she is going through a change of life. Our pods become heavy laden and will sometimes drop off and drift to shore. I would compare it to the musical treat, "The Rite of Spring" by Stravinsky. The climactic point of the music is analogous to having Mars close to us. We feel a frenzy of emotion and a catharsis followed by calm.

We had come to think of Mars as our fatherland, and now we are here on our motherland, which feels much more complete. Much of the culture that existed on Mars, although divided and dissipated on earth, has still been inherited. The difficulties of the "cosmic spiritual warrior" are still being worked out here. We see your evolution of the cosmic mother and father joining to produce a culture which creates offspring that desire a life of beauty and adventure combined.

We can harvest ourselves. Marine biologists believe that tidal action washes us up on shore, but if we wish to remain where we are, we can do so. If there is a need to be on shore to demonstrate our value as a food source, we allow ourselves to be washed up the shore, by the tides of Mother Earth. Wherever we are there are many insects that enjoy us. They are not there to be an annoyance, they are showing you our value as a food source.

Due to the catastrophic pollution humankind has created, such as oil spills, you will not find insects eating off of us as often. They know what is healthy for them and what is not. They show the way, demonstrating where nutrients can be derived. Oil spills are like a plague for us; they greatly affect our home. When other sea creatures and sea plants are damaged by this plague, it is very difficult for us. It is such a tragedy. Those who cause these spills through accident, carelessness, or casual disregard do not understand that we are all

interconnected. Carelessness with your body to the point of acci-
dently chopping off a finger is not dissimilar to what you do with
pollution in our oceanic home. Your neighbors will inadvertently
let their dogs out to make a deposit in your front yard; this is an
annoyance, as opposed to a life-threatening situation. If your
neighbor dropped 50 million gallons of oil on your house, it would
be difficult to go on living, would it not? We are compatible with
humans unless they inadvertently and carelessly dispose of that
which they no longer need into our home.

Many of the species that provide food supplements for you are
"on vacation." The vegetables and fruits here now were adapted
from other environments that were much harsher and where life
was a struggle. Currently, they find themselves in environments
where water is often plentiful. As members of the food chain and
contributors of a nourishing crop, we, along with others, are having
fun. Offering crop after crop is in no way a sacrifice. It is as though
we are sharing the wealth of our surroundings with those who
surround us. We are living a luxurious life compared to what we
have known before in our native climate. Our needs are taken care
of; we are surrounded with variety and offer nourishment even in
our death. When we die, we travel a circuit. The essence of our individual personality immediately goes to a very warm, bright spot.
We are in the center of the earth, the center of Mars, and the center
of the sun simultaneously. This triangular shape has an important
symbolic value to us. When these three points meet on this cosmic
chart of transition, we evolve back into the Creator and await reincarnation where we are needed.

We notice that there has been a change in the frequency of the
earth. We are very cheerful about this. As we experience this change,
we become more aware of variations of our species on this planet.
Before, we were only aware of the other individuals in our immediate group or family. Now, we are aware of the family of consciousness of our species. We do not have a one mind, but seem to be
headed in that direction.

We do not experience sight. We do experience a form of touch that
ebbs and flows. We are aware of our contact with the fluid of the
water and other forms of life. That sense of touch gives us the ability to contact other members of our species physically and to create
a dance of welcome. We have a combined version of taste and smell
that occurs inside the pods. If the fluid changes order, color, texture,
or taste, we will alter our enzymes to achieve balance.

We experience time during growth and inception. Spatial reference is different from your own. Since we are in constant touch
with fluid, space is a continuum. We have a strong spiritual
inspirational sense from the Creator. We feel a sense of kindred

spirit with the dolphins and whales. They live a profound existence and have fun doing it.

The best way to communicate with us is through touch. Those who do deep sea diving can approach one of our beds and see how we are in constant motion or dance. We would love it if you would imitate this dance. It might be very difficult with the equipment you are wearing, but if you could do so just for a few minutes, it would be a wondrous experience. We would know for certain that you are attempting to experience with only us. That is important to us. As adults do not often speak directly to children, you do not usually speak to us with the full recognition that adults give each other. Those who do not swim or cannot reach us in physical proximity, can picture us in their minds eye and attempt to flow and move as we do. Send a picture of your own pleasure in this motion to us. Put yourself near flowing water to stimulate your imagination. If that is inconvenient, sit in a bathtub and make noises in the water. Listen to the noise the water makes and use the sound of the water to have flights of imagination which stimulate your creative energies.

Our auric field is nourished and supported by Mother Earth. The medium of water makes a very good conductor for the electrical element of our auric field. The magnetic aspects are nourished by contact with water as well. In water, we emit a very low frequency sound, and if instruments were allowed to run long enough, they could chart a line drawing our sound. It would be of greater value to use an instrument that does not interpret sound through a listening device, but instead attempts to interpret energy through a gentle connecting device.

It may be difficult for you to understand that a plant that does not stand up and converse with you verbally is an equal. Plants, animals, minerals, and elements are aware of this equality. Only humankind does not understand this. You are attempting to find yourselves and accept the basis of creation. We relate to you as equals and wait patiently for you to do the same with us. Even though your species varies greatly, you can begin to see the simple similarities and not look so strongly at your differences. You are all one human family, and, beyond that, we are all one family in the cosmos.

Chapter XV

Lotus

I am the lotus flower, layered as are the multiple levels of your own consciousness. You peel off layers and grow into new levels of understanding of how you fit into the life of all of those with whom you communicate and interact. Our layers represent the multidimensionality available as a development of expanded consciousness. Our layers of consciousness were perceived by ancient humans, and as we unfold, so their lives unfold. They appreciated the sublime symbolism of our appearance. Before we open we are quite simple looking, and as we unfold, we reveal more of ourselves and reach fulfillment upon maturity.

We act as a stimulant to the chakras, especially the crown chakra. Within our own bodies there is a single chakra, closely related to your own crown chakra. We are also aware of a little-known chakra, located between the knees. This is related to the corruptible nature of humankind, and we do not associate ourselves with this chakra.

It is possible to be near enough to a lotus to cause the auric fields between us to intermingle. Our auric field is primarily gold, with a small amount of purple and white light. Utilizing the power of concentration and an actual visual memory of the lotus, one might stimulate communication. This is more useful than a photograph in connecting with us. It is not necessary to touch the plant physically, but for those that do, or for those that are sight impaired, touching with the back of the hand is preferred.

We originated within the planet Venus and were created by a race of feminine beings. They created a plant that symbolically interacts with all levels of beings who then achieve a greater understanding of their own interaction with their worlds. We are a spiritual as well as a psychological plant. We derive from a planet that is an archetypical symbol which represents love and passion. On the emotional level we, too, are associated with love and passion through the balanced flowering of femininity and masculinity.

The surface of Venus does not lend itself to flowers. Inside the planet, at a quicker frequency, physical conditions encourage our life style. We are much larger there. If you could see us in that dimension, the size would approximate a ten-story office building. We are not a single entity; we are more associated with all forms of my species. We equally identify with the ten story tall versions of ourselves, as with the tiniest miniature version of the lotus which exists under the water on Sirius. Due to their moisture and humidity, we easily adapted to many of the water planets on Sirius. We arrived on earth indirectly, by way of expeditions from Sirius.

We experience forms of ourselves, other lotus variations, wherever they might exist. We are consumed as food in our underwater lives on Sirius by various aquatic beings. We are used as a form of mental stimulus on Andromeda, much as you seek aphrodisiacs on this planet to stimulate sexual powers. We are used on Orion to stimulate visions through use as a massage oil.

There are aromatic oils that can be created by mixing us with other flowers, particularly sunflower petals. These can be useful in raising consciousness. A small amount placed upon the third eye area, the solar plexus, and the heart, used during meditation, can stimulate an expanding consciousness. Please use respect in the gathering of these petals. Portions that can be gathered need not be from the live portion of the plant. There are portions of the plant that will die off. If you must harvest from the live versions of ourselves, please only take a small portion of the plants in any given group.

When we are no longer physical, we return to our normal state of being. Our etheric bodies return to Venus to be recycled as energy and reincarnated as other forms. We tune in to the center of the planet, Venus, for the source of feminine energy.

There is a blending of consciousness of all forms of our species, and when we choose to have a group connection, it is through the dream process. We do not dream about situations that are not associated with us in some way. We have strong bonding occurring then and receive any necessary messages. In the dream state, we have spiritual and emotional communication. We have so many other species living in other dimensional aspects, that we feel stronger in our multidimensional aspect. When we are in contact with extensions or variations of ourselves, we feel ourselves as a complete unit. Any single plant may not feel fully complete without the ability to contact other members of its own species.

We feel that when we are surrounded by life, the Great Deity is within all aspects. We feel more connected to the Creator in our conscious state as compared to our dream state. We are not here to face problems, as the human race is. Neither are we here to create difficulties which we can later solve. We are here primarily for a life cycle, to serve as a symbol, and for the pleasure of existence. We feel very balanced in our completed harmonic self. We share with you during meditation a form of our conscious harmony. We have much to offer in analogy to the unfolding of human consciousness.

Those who meditate with or on us may feel a similarity in consciousness. We have achieved a sense of peace and acceptance in our Divinity. We are able to impart this through our radiated energy. Many other plants can do the same but are not as strong a symbol.

The awareness of our symbolic ties to humankind's evolution forms a mental body. We have a spirit and know that all is connected to the Divine. We have a form of emotion which reacts to our immediate environment. Our primary emotion is not fear based, but rather based upon a feeling of familiarity. We are most familiar and accepting of our own species. Those who achieve lotus consciousness in your human race are also included in our feelings of familiarity.

We do not feel the earth's shifts and changes in dimensions as strongly as you. Your life styles are speeding up and you do not have as much time to experience situations or to examine them as slowly as you have in the past. We have, by nature of our being, a great deal of time to explore our immediate environment. We do experience life as a simultaneity of events. We are philosophical in our reaction to humans. We feel the vibration of your voice, and if we hear words spoken in anger or violence, we feel threatened. We will seek connection on a stronger level with our own kind to feel continuity. Continuity is a wonderful energy to sooth the feeling of loss.

Time and space in your dimension are inevitably tied together. In the third dimension our spatial reference is greatly modified. We do not experience space other than an individual plant's reaction to that which comes close to it physically. We do not experience time other than in our reactions to day and night.

As you reach for your highest expression, notice that in this day and age you are surrounded by temptations. Know that they may move you off of your path. Be clear about your choices in life and how you want to feel and interact with others. Even though you are surrounded by technologies which are sometimes constructive and sometimes destructive, know that the human race has been artificially stimulated to achieve the highest levels of communication, survivability, and exploration. You will achieve an age when all machines do not have moving parts. The great re-crystalization stage will be the final step of coming into your true God-ship.

Chapter XVI

Marijuana

I am speaking to you now as an individual plant, but I connect with the larger aspect of marijuana's mass consciousness whenever I choose. I am growing in a pot, as one might grow a flower, in the home of a law enforcement individual in southern Colorado. I am not being grown for purposes of distribution. My primary use is for demonstration to groups of children. I speak to you at this time due to the misuse of myself and my species.

Our origination point is from a distant planet in the Andromeda star system. We were once a large plant, a bush growing to heights similar to that of a small tree. At that size we were very durable and yet considered to be unadaptable to earth since it was our propensity to grow in profusion. A miniature version was created by cross-breeding us with a grass from the earth. Were you to smoke a version of our original selves you would receive clarity and insight on the first puff.

We exist on other planets where there is a desire for enhancement of spiritual qualities rather than for the purpose of dulling the mind or senses. This is not to suggest that other materials are not used for this purpose, but I am not aware of our misuse. Plants related to us in their effects are used as liquor-type intoxicants. Their effects are greatly diluted in comparison to your experiences here.

When dried out for smoking purposes, the bulk of our physical structures no longer feel pain. There is an energetic aspect around the dried product and around the smoke as well. When we are burned, we feel transformation. When burned alive, we feel pain. This is why we are compatible with you in your spiritual quest. Our emotional bodies are not so unlike your own. After we die we return in our physical structures to the earth. We support our continuance through our seeding procedures, and on the soul level, we immediately return to our point of origin on Andromeda where we are energized by our true form. We find a bush and interact with our natural selves and experience the Creator. We are refreshed and are given the choice to stay on Andromeda or to return to earth through reincarnation into a physical plant.

Marijuana has resided on this planet for quite some time. We are here to assist you in perceiving the value of your spiritual alignment and harmony with the earth. Originally, we were intended for use as a medicinal herb by members of the priest class. Marijuana is currently used by shamanic and ritualistic people to enhance the visionary and compassionate qualities in the user. We are gathered from the wild, where we chose to grow naturally. Individuals taught by shamans or those trained in various medicine paths know that the plant itself is never fully harvested. The leaves are collected, but the plant is never destroyed. If it does die out, it is only temporary. Those who are on the medicine paths and trails, those who have been initiated into understanding how plants, roots, minerals, and other elements on earth can be used to enhance spiritual growth, know which plants to use. They are drawn to a specific plant.

We have a name for ourselves which is not pronounceable in your vocabulary. It sounds like a song that carries on for a long time. Individuals who desire a particular plant can find us through song. Through imagination, quiet song, and gentle tunes, you can communicate with us. This would be appreciated. We enjoy classical music. You can hum or sing a tune with whatever words you choose, and when you get to ranges that you cannot reach out loud, imagine what those tones would sound like. We can interact with that. We know when you are reaching out to us.

It is intended that you use us wisely and sparingly, even though the temptation to smoke us as a method of escape is strong. If you use us as an escape from your problems, you become dependent on escape. We are a challenge to you. We see that temptation as an aspect of your own dark side and consider improper usage to be an abuse of power. When misused we represent the mirage, that which only appears to be there and appears to give one hope in the struggles of daily life.

When we are abused through greater consumption than was intended, an element of our structure acts as an enzyme to stimulate hormones. This stimulation becomes addictive and is referred to as a psychological addiction, but it is more seductive than that. It is a condition that creates a bridge between mental addiction and physical addiction. This makes us a controversial plant in that those who wish to eradicate us perceive us as a drug used to enslave populations through their dependence. We understand this attitude.

We work energetically to stimulate brain chemistry, which in turn stimulates olfactory memory. The memory of smell utilized for spiritual enhancement may be the safest usage in a society that is struggling with addiction and abuse.

I am not trying to convince anyone to become a drug user, but if you do begin using us, do so with the idea of solving problems and requesting visions. We see ourselves as support in allowing you to access levels of your own wisdom. We provide information, advice, and solutions to problems, both on-going and short term. The appropriate way to use us is without paper, that is to say, smoke us in a pipe. If you do so, one puff per hour or hour and a half is enough.

We are designed to stimulate your mind, not to suppress or dull it. Marijuana is designed to support that which you already have. It tends to break down the barriers that you build with your ego in order to survive in a world of struggle. It returns you to a euphoric state of being that is associated with very early childhood or with senility. It slows your perceptions down so that you can perceive the subtle. It is within the subtle and simple motions of reality that the solutions to your problems are available.

We will be recognized, at some point in the future, for our ability to enhance mental or visionary powers. The sparing use of us will help to assist those who do not have visionary qualities naturally or do not cultivate them. Your race will come to depend on the visionaries in your society. Very young people will be the source of this visionary ability, not only with their increased mental powers, but with their ability to use their imagination in ways that are applicable in your scientific and technological communities. We will be recognized and researched by Eastern scientific researchers who are less hampered by demands to produce profit.

In the West there is an attempt to synthesize certain elements from us, to use us as a substance to support pain relief in critically ill patients. Creating a synthetic element from us does not support our personal desire. We prefer to be used in our natural form. Medicinal properties have been discovered in association with space travel. This research is unavailable due to its connection with highly secret studies that are associated with developmental technological projects. Due to political pressures, it is unlikely that we will be used in long space flights. It is conceivable that substances derived from us may be used to alleviate boredom in the future.

We are aware of time and space. We are aware of time in the respect that we observe the day and night cycles. We enjoy the warmth of the sun and coolness of the evening. Our dream state and waking state are not separated. We experience space as a continuum. We know that which is immediately close to us; if a creature passes by, we are aurically aware of its interactions with us. In the wilderness, most creatures have an instinctual knowledge about us and do not consume us regularly. They do consume us instinctively if they have some form of gastrointestinal discomfort. Rabbits are one of the most spiritually aware creatures who occasionally nibble at us.

We do not communicate through thought directly. If it is necessary for thought transference we go back to our source. If we wish to contact a member of our species, we use our auric field to project as well as to receive. We communicate through sound in ranges which are beyond the capacity of the human being to hear. It is a very high ultrasonic wave.

We are similar to you in the sense that we have the desire to survive, and we have the desire to serve towards the higher aspirations of spirituality. We procreate with these ideas in mind. We want your acceptance of our right to live, but if you must wipe us out, say a little prayer over us. Do this as you would over any beloved form of existence. This will help us feel that you valued our life, as you value the sanctity of all life. Learn to appreciate your own true purpose. Every one of you can come into a greater understanding of your own spirituality and can regain your original power.

The Creator of All That Is requested that certain souls emigrate to planet earth. Here you experience the cycles of joy, pain, suffering, and happiness. Many highly sensitive souls incarnated as priests or priestesses. All of you, in those incarnations, were anointed with a quality of self-sacrifice. The anointment refers to a destiny, a path that beckons, a search, a quest for the Creator and "home" in oneself. It is a touch by the Creator on the forehead of humankind, that allows them to rise above the depths of depravity and perform the most noble of acts. It is a responsibility as well as a gift.

The struggles you find yourselves involved in are beginning to threaten your actual existence on this planet. You make sacrifices, including your own lives, for the perpetuation of your culture or of your immediate family and friends. There are children being born now who are even more sensitive than you. They need advice and guidance to get along in this world while maintaining their sensitivities. Your struggles need not be in vain. Recognize that even though you struggle through dependencies on alcohol, tobacco, or drugs, those of you who live to tell about it will help to guide the young ones. They are directly involved in reharmonizing your world.

Chapter XVII

Opium

Greetings, I am speaking to you as the opium plant. I speak from a place of universal resonance. We represent seduction and warmth. Opium is to be used as an aroma or scent, not as a chemical invasive agent. We are sought after by those who use us for life encouragement as well as for purposes of disruption of the natural flow of the body chemistry.

Our point of origin is in the fairy kingdom of this planet. Due to humankind's lack of conscious awareness, this kingdom is not encouraged to exist. The kingdom exists beneath the surface of the planet, in caves and shelters. The fairies can function in the second, third, and fourth dimensions. When they do not wish to be seen or heard, they appear in the second dimension only. If they turn to the side, they seem to disappear. Scent is very important to the life of a fairy.

The Devic kingdom and the fairy kingdom are deeply interwoven. There is a uniqueness to each and there is a family connection. Our Devic energy is both masculine and feminine, strong and gentle. It is a polarized form of Devic energy. This is an exceptional level of Devic energy, as most energy of this nature is known for encouragement and sustenance. Some energy can be demanding and bordering on stubbornness. When the time is appropriate, this energy can be aggressive. Our energy is both gentle and aggressive. This is a valuable reflection of humankind in the Devic Kingdom.

The fairies know that our usage is designed to enhance life. If we are used to alter life, we do not become angry or defensive, yet we are aware that misuse can lead to self-destruction. Our intention is to confront the seductive elements in your own society and create, in time, a greater understanding of warmth, love, and sustenance.

Our effect was originally intended to stimulate imagination and the childlike elements of imagination in all beings. In your culture, you believe that once you move beyond childhood, your responsibilities change. It is not necessary to fulfill in the physical plane what exists in your imagination. It is, however, necessary to maintain and encourage it. Valuable ideas spring from the imagination.

The best way to communicate with us is to use your childlike imagination, which we stimulate through our scent. You can ask yourself what a plant would say in response to your communication. Dance or physically express the joy of our presence through imitative actions. These games will stimulate your imagination. Benevolent interactions between humans and plants is beneficial.

The purpose of all life on this planet is to experience the ultimate balance in the physical world. The extract heroin tends to keep people from physical experience or to so grossly alter it that they do not experience actual interaction between themselves and their physical environment. They do experience an altered state, though these states can be achieved through a subtler meditative or prayerful state of consciousness. Humankind's impatience drives them to reach levels not unlike your perception of Nirvana. In Western cultures, your idea of Nirvana does not always equal the actual state as it is achieved in the East. The intention of Nirvana is to achieve a fully activated spiritual state while in the physical and emotional world. The rushed idea of creating an altered state through the use of heroine or other such substances, creates a condition that is suppressing your perception of the physical world.

The dual side of our nature is designed to bring forth an understanding of your own unique possibilities. You have your light and dark sides. We are designed the same way. We represent seduction and warmth. An aspect of friendship could be a desire to seduce and make the other person dependent or needy. The aspect of need describes us as we are used in drug form. The best way to transform your own desire to manipulate, to create a sense of desperate need in others, is to use us in the more positive aspect. We are uniquely archetypal to the challenge for human beings now and influence the full range of the human condition. We are involved with the human destructive nature as well as with the constructive nature. It is not possible for us to choose to cease to exist so that you would not misuse us.

There is a similarity in your experience with us and our own experience of ourselves. There is a communing with the past and future of the individual, who are under our influence. This is sometimes a pleasant experience and other times not. We do not advocate our misuse as a drug. You exist here to pursue a physical evolution of yourselves, and these extracts tend to decrease your physical activity.

If you wish to reach your objectives through destructive means, it is allowed. The difference between allowance and encouragement is that in allowance you proceed under your own power, in encouragement you proceed under your own power and Mother Nature's influence. When you go downhill on your own force, you can use the laws of gravity, but when you go uphill, Mother Nature will give you hints of the value of this direction.

When we are cultivated by individuals whose intention is to use us as a disruptive chemical, we feel resistant to cultivation and are not always cooperative in this growth. Those who grow us with a creative and benign intention find us growing more bountifully. We understand the need and value of human/plant interaction and accept this with grace.

Forgiveness begins with the idea of allowance. In the plant kingdom, we allow that other life forms exist. When growing in the wild, if a creature such as a cow inadvertently steps on us, we do not feel angry for destroying or injuring us. We allow that the creature, as a survival need, stepped on us. The basis of forgiveness is the understanding that all creatures and plants have a right to pursue survival. If you can give yourself the right to survive and pursue life, it will greatly enhance your ability to forgive yourself.

The human being must develop a certain degree of latitude and flexibility. Your souls exist beyond the physical world. Your souls have a desire to experience all ways of living in the third dimensional physical world. Sometimes you pursue things that are destructive to yourself or others. Realize that whatever you pursue, it is done with the constant search for self-identity in the physical. When discovering the self in the physical, one always has the example of consequences. If you can forgive yourself for the destructive consequences that result, you will begin to understand the core of the idea of allowance.

We, as fulfilled beings, understand Mother Nature's ultimate allowance and experience the total vertical connection through our dream time selves and experience her range multidimensionally. Our relationship with her is most intimate. We experience spirit as our "now time essence" of the nature of our full personality.

Dreaming, as you understand it, is our closest tie with you. When you dream, you have many levels of the experience. When you awake, it may feel like a linear experience, but in reality it is a multileveled, multifaceted experience of yourself, some of which you remember, some of which you do not. We are in the experience as a plant all of the time. Our constant involvement in the multilevels of multidensity experience includes dream time. There are some animals that exist in dream time as well as in physical time, such as the dolphins. We exist in this time, all of the time.

We have a vertical state of existence in which we experience all of the levels of ourselves. We experience all of our teachers as variations of ourselves without flowers. The spirit versions that we work with the most are those that exist in the distant past and have very strong ties to the distant future. Our constant state, although physical, is also non-physical. We are in 24-hour physical and dream time simultaneously. This may account for the altered state experience of those who use our extract. Our original purpose for being used as a scent will create, stimulate, support, and sustain an imaginative state that is referred to as hallucination or daydream.

We have a mental body; this is why we are able to experience multidimensionality as a visual experience. We do not communicate as humans do through language, we do it through knowing. This is the realm of knowledge that exists always. We do not communicate about the events of the day individually, rather we know.

Our relationship to time, is multilayered, and not as strong as your own. We can experience linear time, though we are more in touch with now time. Time is basically experienced through our life cycle. This gives us the opportunity to understand needs, values, and philosophies of our own species. This is given to us to understand the interaction of the human being with us, which is prevalent these days.

We experience our environment through the sense of auric touch, as well as through the sense of physical touch. The human being has these abilities as well. We do not see, we sense through a harmonic vibration, which is a touch/tone scent of all that we come into contact with.

It is important to use the scent of the opium while the plant is in flower. The best way to experience it is while it is living. Even though well-meaning people create extracts of plants, concentrating them highly with the intention of speeding the process of stimulation of the original scent, it is not our intention to speed up anything. Rather we encourage, through the harmonic vibration of scent, a gradual evolution that stimulates the idea of sustenance. This is the nature of the balance of Mother Nature on this planet. If one cannot obtain us in our natural state, as is the case for most people in the Western world, the next best thing is to obtain the dried flowers, rub them on your arms, legs, or hands and then smell the fragrance.

If we are used in the form of "flower essences," the scent can be used in an act of love, not necessarily sexual love. It can mean kindness, friendliness, or warmth. It is necessary to be involved in these emotions to use it properly. If it were done in a distracted way, or if the emotion that was present was anger, it is possible that the extract will stimulate seduction rather than warmth. One must be aware of how to use it, as well as what can be done with it.

Mother Earth creates balance. The human being needs to learn to imitate this balance. Humans are not in charge of maintaining our balance, they are here to learn our balance. Currently, you do not represent earth, you represent tenants and caretakers of earth.

There will be an increased awareness of the plant life forms that are gradually leaving your planet, due to your destructive nature, "I want what I want when I want it." We will begin to be used as we were intended. The expansion of human consciousness will be preceded by a contraction of consciousness. You now see your world in a compressed, encapsulated version, in order to understand the nature of survival.

When humans or plants do not feel welcomed by those they are setting examples for, they will go away. If individuals do not feel welcome in their community, they will not stay. If people do not wish to imitate what you are doing, you may change what you are doing, or you will go somewhere else to be appreciated. As plants, we leave seeds to germinate in the soil. We only return when there is a commitment by the human being for a sustainable balanced culture. Then it is possible for the regeneration of plant and animal species. A sufficient amount of time must go by in order to make us feel welcome. Prehistoric plants will regenerate if there is sufficient scientific curiosity, hope, and desire. Desire can be created through imagination and dreams. This is why we encourage imagination.

Humankind does not yet fully grasp the growth and recycling experiences occurring as you come together to support and sustain the continuity of your planet through challenges on the pollution level. You will interact more with other cultures and realize the essential element of forgiveness in order to go forward. It will not be possible to isolate governments and say that they are the bad guys and bump them off. In order to rise to the most immediate challenges, such as pure water and air, you will become dependent upon each other. In Mother Nature's world, all plants and animals are interdependent for sustenance and balance. You are approaching an understanding of maintaining balance through dependence upon other human beings, on a polarized, negative need level. However, since this is available, it is considered to be better than nothing. Once you rally around the cause of crisis pollution and find ways to work together, you will eventually recognize the value of cooperation. When you need each other absolutely to create permanent sustenance of life, you will recognize your interdependence. As your numbers increase, you no longer have the luxury of affording wars, due to the level of destructive devices that are available to you. You will learn how to rebuild and your armies will become the armies of rebirth rather than of destruction. This is possible. The first world order will come together to create sustaining government and economic cultures. This initial crisis will show what works in a world order and what does not work. It will take several years to create a sufficient backlog of reflective knowledge in order to truly function as a support system for something new.

You have utilized means of cooperation in the past that have not sustained themselves. Civilizations have come and gone. It is not the intention of Mother Nature to suppress individual cultures and uniqueness. You must learn how to work together through crisis. In the past, there has been much destruction in the pursuit of individual wars and the soul of Mother Earth cries out for solution! Your own souls cry out to be delivered from their situation of need and to create a permanently sustaining harmonic society that can imitate Mother Nature at her best. The driving force of human nature is survival. Mother Nature has allowed the pollution crisis even though there are destructive elements of humans' darker nature working against her.

The ego is humankind's attempt to allow the mental body to interact with the physical world. It is of value for you not to abandon the ego as something that is primarily manipulative or destructive. The plant ego is that element of ourselves that gives us the precognitive ability to sense our own imminent destruction. The ego functions equally on the electrical level, to give us the sense of where to grow in the wild, where we are best sustained by Mother Earth. While destructive ego appears to be that which creates disharmony, it is also in constant search to define itself. That which you do destructively creates immediate problems. However the long range effect is to create a negative definition of that which does sustain. If one does not immediately gravitate toward balance and one chooses to define balance by going everywhere but balance, one will necessarily deplete the choices of negative expression, thus forcing oneself to observe a more sustainable positive expression. Negative ego is simply representative of choice.

When the entire plant species is under threat, certain plants become stronger, like those that have the capacity to create dependence and desirability. The plant kingdom is under siege. Everywhere you look the soil is quickly becoming desert land. Fertile soil is becoming endangered.

In the plant world, there are several plants designed to be "soldiers or protectors." We are not soldiers in war with the human beings, but when opium is bombed with aerial sprays to eradicate us, we come back even stronger. We have created durability for ourselves, as the marijuana plant did. We thrive in the wild and grow even stronger when encouraged to mature.

When we are no longer in a physical form, we experience the multidimensional levels of spiritual representation of ourselves. During our time of not being physical on this planet, we experience stronger focus in the other realms. We have a secure sense of identity with all of creation. This is instantaneous. The sense of continuity is very present. This is how we describe death as you know it. We do not feel separation but a different focus of our continuity.

Know that attempts to join with the Creator through mystical alterations of your physical experience delay you from joining with the Creator in true fulfillment of the human purpose. To be fully physical is to be fully physical. Your lives are constantly challenging, and yet, in order to become the ultimate of your Divine purpose, challenge becomes your ultimate pleasure and solution. Challenge, in this case, represents the masculine physical principle. Solution represents the feminine principle. Embracing the two represents total earth harmony.

Chapter XVIII

Pine Tree

I am speaking to you as an individual who is living on a hill in the Angeles Crest National Forest. I am a particular species that you admire and are reassured by my peaceful existence in apparently hostile country. I am a member of the pine tree family most often seen growing out of craggy cliffs and rugged places, as if springing from the rocks.

I enjoy, through the union of all tree consciousness, the "emotion of completion." I feel like an individual, yet when I join with others of my species, I feel like a larger portion of myself. I am cognizant of a feeling of a group soul and in that light I will speak of my kind as a "we."

We have an understanding with the rocks, and we conduct a form of religious rite that is necessary in creating our root structure. We ask a rock which already has a crevice to allow us to pass through it. During our growth we will sometimes expand the crevice. We communicate with the rock as we pass through it and receive a form of permission. This could be perceived of as not being necessary on your planet, however, we are a formal species. We like being asked and informed if things are going to change. Equally, we like to inform.

We are oftentimes appreciated for our ability to "eke out a living," that is, survive in the most inhospitable conditions. Very specifically, what I represent for the human being is strength through struggle. As human beings you must often struggle, just to stay alive or to keep what you want. Very often this struggle will make you stronger.

When one has a long life as we do, provided there is no overt disturbance to us, we have much time to observe our own interactions with our world, your interactions, and the interactions of various animal species. From time to time, if there is going to be an event which may cause us to be destroyed or changed, we will have a precognitive dream of this to forewarn us.

We will very often dream of our home planet. When we run down our tap root on that desert-like planet, it goes down to the liquid which sustains us on that planet, like a lifeblood. It is as if we were tapping a vein or an artery for ourselves which fills us with more than the nutrients of water. It fills us with all that we need to be completely fulfilled.

We often dream of that energy of fulfillment, and then we share the knowledge of our dreams with other trees elsewhere. It is not as if we lean over and whisper into each other's ears, but the knowledge of the trees does slowly go around the world, and we learn about that which other trees have experienced elsewhere. We are distantly related, oddly enough, to the redwoods. The redwoods, as they have existed in the past, did not have red colored wood. Our chain of life would not find us related to trees which exist on your planet now, but we are related to trees that existed on your planet in the distant past. Most of our relations, "cousins," passed on in the prehistoric time.

We originated in quite a small desert environment. It appears, tapping into my source consciousness, to be a planet that is not unlike Mars on its surface, but is in a distant galaxy far from here. As close as I can approximate, the name of the place is the Pyroxides galaxy. This galaxy is many, many light years from you. The planet is very dry and has underground springs, with very little rain, as you understand it. Once or twice a month, in terms of your time, it has a heavy dew. We do get some external moisture. On our home planet, we have roots that go down for many miles. We have the ability to reach down very far to obtain that which we need to survive. Just as you very often will reach down deep inside yourself for strength.

We have existed on Mars in the past. The energy of Mars required one to be strong due to it's nature and temperament as a planet. There will be reference to that idea when your explorers begin to distribute information about what was found on the surface of Mars. In the future, they will find a form of pictograph illustrating us, as well as perhaps some other form of documentation.

As far as our source in your solar system, there are no other planets of origin. We do exist on other planets where dry habitat occurs. We often exist on a planet previous to the beings that will inhabit that planet. We prepare the planet to understand the basic requirements of the beings who will come to occupy it.

We find that we do have a relationship with your Moon. As the Moon moves towards her full cycle, we experience ourselves more fully as well. This has to do with our emotional bodies. I personally find that I become more sensitive. My ability to feel a passing bird or insect becomes greatly heightened. I do not feel quite as strong in my abilities when the new Moon is present. If another planet, say Mars for example, rises within my awareness and passes over my location, I feel that there is some change, although it is very subtle.

Those changes primarily affect human beings, who then pass those changes onto us by their actions.

The primary sensitivity that we share with you is the sense of touch. We are very sensitive to touch. You may be surprised, but we actually know when a fly lands on one of our leaves or branches. It is a knowledge that is useful to us.

When we reach out into the world to broadcast messages to others of our species, our auric field will get very large and very sensitive. We will actually feel the fly heading in to land on us.

We do not move around as you do, so we experience time and space at a much slower rate and as a continual unfolding. Our world is smaller than yours. For example, if a fly is on one branch and a bird is on another, we experience the spatial reference in terms of the distance between the fly and the bird. It is a small space, and yet when we expand our consciousness to connect with others of our species, we become increasingly aware of one large mind-emotional body.

Emotions are some sort of universal translation experience which works with all beings who have an emotional body. The primary communication from one species to another is understood or interpreted by this form of translating ability. We communicate through telepathy, or vibration communication to other plant, mineral, and animal kingdoms. They also have the ability to understand our vibration and what we have to say. We use a slightly different vibration for each.

Actually all of our perceptions are based on physical and emotional feelings that are totally married or bonded together. We are familiar with the concept of colors, though you experience colors through visual input, we do not. We create a perception of color based upon how we feel physically and emotionally about an object. It is uncertain, whether we would see the same color. My experience of color would be based on my imagination (my ability to create an image in our group mind, or my portion of that mind). We do perceive a form of moving image that looks like an electrical picture. It is not as in a TV screen, but almost as though the electricity is alive and moving.

Different colors represent different values or densities of that electrical image. Perhaps the best way to describe it would be like seeing lightening close up. Those are the types of colors that we sense. We have been told over the years by many that we radiate a healing energy. We have come to believe (in our group mind), that it is our purpose to be involved in this healing through our emotional body. The emotional body is such a large portion of our active consciousness and is radiance itself. When you are struck by a radiating body of energy that is keyed to emotion, it amplifies and increases your own energy connection to your soul. This must, in it's own right, stimulate your healing powers.

We are very sensitive to air pressure. Your music creates a resonate air pressure and a tone based upon the movement of air. We are more sensitive to the air pressure on the emotional level. We notice the change in the mood that is around us before a storm is present. While you have the ability to sense by smell the electrical scents in the air in an electrical storm, we feel the change of pressure.

There is a sound that we find appealing and I find individually attractive. I can attempt to spell it phonetically by saying "Foesss." I am told by some members of our species that there are certain groups in Tibet and other areas of ancient China who have developed sounds that are used to heal the human being. That sound which occurs in nature has some healing aspect for myself

For those interested in aromatic essences, like holy oils, you will understand our contribution toward their arts. We will use our scent, which we find particularly alluring, to attract the fauna involved in our process of survival. Human beings have specific scents associated with themselves as well. From our perspective, this appears to be quite normal to life.

The uses of our trees for spiritual healing and meditation purposes have not been fully understood by many people. The teas and the balms are the best remedies to use for medicinal purposes. Please check Native American literature for the teas that they brew containing parts of us. Of all the trees that you could lean against, meditate upon, or pray near, our species is one of the best. We feel that our spirit selves are in constant harmony with our physical selves, and we thus support other life forms expressing themselves the same way.

If you wish to communicate with us, I ask you to call forth an emotion that would match what you wish to say out loud. Since it is normal for you to communicate in the spoken word, you can then be harmonizing your emotional body with your mental body. You notice that a great deal of energy goes toward this type of communication, since normally you just use the spoken word alone.

Those who are photographing us using Kirlian techniques (energy fields) have noticed that we respond considerably. I can assure you that if someone cuts off one of our limbs we will recoil in our emotional body. Since our energy fields, when viewed visually are largely the equivalent of our emotional bodies; our reactions can be seen. We do appreciate formal warnings should someone cut us down.

If you take a match to a plant or cut down a tree without first telling it that the action is needed or required, there will be a period of mourning. During this time, we go inside ourselves to the source of our strength. We do this not to hide, but rather to seek the maximum strength we have within ourselves. We can also be amused. When we are amused and feel particularly joyful about our world, our energy bodies will radiate way, way out and can be perceived by those sensitive enough to perceive them.

Environmentally, we have the greatest concern about what you call the "acid rain." We understand that it is not a direct attack upon us, but we do feel unwanted. We feel that you do not understand how we contribute so valuably to the lives of many species, as well as your own. We do feel unappreciated. This is why your scientists have difficulty with their predictions and the actual outcomes of your air and water "pollutants." Depending upon how the trees feel in any specific area, an entire forest of a particular species can die out. They will do so strictly on the basis of feeling less and less respected by your people. We will live longer if we are enjoyed because we feel that we are needed.

I understand, from my grasp of your future (which is something inspirational), that you will create a raised roadway system. You have already begun this with some highway and freeway systems. Eventually your transportation systems will not utilize so much of the surface roads of the planet and more of the natural plant life that occurs on the surface can remain. We are looking forward to that day. In the meantime, when we die, we experience the return to our Source. We return by moving down into our tap root. We move down into the area of strength and then pass on our knowledge to others of our species to be added to the library of knowledge of our species. We move, after this period of grace in our tap root, into the skies and the heavens and go on to our home planet of origin. There we bask in the rich warmth of our Creator. We return when we are called upon to spring forth into life on some other planet or again on this one. If it is perceived that we contribute in some way, we will again return.

Trees have a parenting instinct for their young. We understand that all of our offspring are portions of ourselves. When the seed leaves the tree, we wish our offspring well as we are aware that the seeds will be taken elsewhere. In the case of an apple tree, it can drop its fruit and another tree can grow up right next to the parent tree. It is a blessing and we accept it as such. If left undisturbed we live for a rather long time.

We feel so very fond of the third dimension which we all occupy. We do not feel any sense of one dimension being better then any other. We are hopeful that you continue to let pine trees serve you. We are an example of what you can attain if only you search, quest, and find your own niche. Run down your own roots in the most compatible way possible with all life forms around you. Of all the species that we are aware of, your species is most capable of doing just that.

Trees have power centers, or chakras, within their form. The tap root is the source and the strength. That which exists above the ground is the dance of life, in which we feel a state of grace most of the time. We enjoy the pleasure of the search. We quest for the place to put our roots, and we celebrate our victory in finding it.

Chapter XIX

Rice

I am the spokesperson for rice, the foodstuff, and speak now for rice at large. My species, occurring in many different varieties in many different countries, feels most connected with the etheric realms. We are not from any particular place. We are spirit materialized into matter. We are not aware of our existence on any other planet. We are essentially spirit in nature.

The reason we have been encouraged to materialize is to provide a food substance to others who are attempting to develop the ability to materialize into spirit form. This is why we are such a popular form of nourishment to cultures which have deeply rooted spiritual and religious traditions. Our homeland is the etheric realms associated with the angelic kingdom, as well as our connections to Devic and elemental aspects of earth. Our auric field connects us deeply to the earth.

We act as a constant. In chemistry, there is a certain constant chemical that is added to allow other chemicals to combine in a catalytic reaction. We are a substance that not only feeds you but allows the Devic kingdom a portal. Wherever we are grown, Devic energy can cycle through these areas. There are Devic windows that are associated with us that may or may not have to do with rice as a substance. These Devic energies will retire or return to the angelic realm through windows near fields of rice. Plant or animal species can come through the windows when they are being re-energized. Even those that are new to your planet arrive through windows associated with our presence.

We feel the emotions of the Devic energies that pass through our windows. It is not that the Devic energies, of their own right, experience these emotions. If a plant species has been retired from the planet, it will often feel sadness associated with the desire for the life form to continue. Their lust for life has not been quenched. It is possible to access these feelings from human beings around us, due to the vertical nature of our consciousness. We pick these up as the elemental or Devic energies passing through us towards a purification in the angelic kingdom. We also experience those who are being returned or re-energized or who are coming to earth for the first time. The angelic kingdom has a constant state of good humor, and we have that basic energy emotion as a usual state of being.

When there were wars in the Far East areas referred to as Indochina, Vietnam, Cambodia, and Thailand, the fields were destroyed through acts of violent conflict. Spiritual energy, though it was greatly needed by the indigenous populations, was somewhat restricted. The people cried out for help, but because of the destruction of the rice paddies, Devic windows were lost for a time. Certain forces that desire the restriction of spiritual energy on your planet have attempted to eliminate us. This is why we are considered by many in the current priest class in Tibet, South America, or Indochina to be a holy crop. They understand the link between rice and the spirit world.

As a crop, in places that are feminine in their geomancy or areas within a tropical climate, we accompany other forms of spirit energy to earth. We do not feel an affinity with other earth plants, as they have been brought here by well-intentioned cultures, often from other planetary systems. We are not suggesting that we are the food of the gods. We were, however, placed here by the gentle angelic spirit energies for the purpose of fostering and nurturing the spiritual and emotional bodys of the human being.

We are here in your lives now to suggest, especially in countries in which we are not a major portion of the diet, that we add a feminine, nurturing quality to your life. In the Western world, very few people consume us in quantity. We recommend that few, if any, chemicals be used in our growth cycle and that we not be heavily processed. We do not grow as many other plants do. We often grow wild, in a feminine water environment. Our symbolic interaction with the human being assists you in understanding that the spiritual self can be consumed and released in such a way as to feed the physical body, stimulate the mind, and nourish the emotions.

We live in higher dimensions while our roots, in terms of our conscious selves, are grounded in the form that we take as we grow by those who farm, collect, and gather us. We are nourished physically by the soil, water, and nutrients. Imagine a picture of spirit rice, living with the angels, with a long, trailing, spindly body. When it reaches the earth in a field of rice, it materializes into matter. In terms of a plant, we may be the least familiar with daily life here on earth.

We intend to act as a bridge substance that allows you to expand your consciousness and gives you guidance, nourishment, and love as you transcend the material plane and return to the spirit plane. This is why certain ceremonial versions of rice are consumed as the body gives way and returns to its mortal beginnings, and the spirit ebbs towards its final reward. This explains why this food stuff is found, from time to time, in burial crypts; it was placed there to encourage the loved one to find his or her way back home. In time, there needs to be an understanding of the stalk associated with the plant, as it has certain medicinal properties associated with its aromatic function. If it is dried and burned with sage there is an expansion of consciousness available to you. This ancient temple ceremony has been used to speed departing spirits on their way.

We provide an energy that allows people to go home. We have been here from the beginning, even before your population was developed in its current manifestation of the human races. We were here before the pioneer races arrived in order to determine which life forms would survive. Rice was given to this planet in the form of wild rice. The reincarnation cycle of your souls, through this planet, are supported by our very presence, regardless of whether we are consumed or not.

We are in constant communication with all spirit energy. We hear through vibrations and we sense changes in temperature. Each of these is a form of communication to us. We are sensitive to sound and particular sounds cause us to relax. We are aware through telepathic energy and have a sensitivity to the vibrational harmonics. This idea of vibration is not unknown to your computer scientists who refer to data as a "pulse." Pulse can be referred to harmonically or musically as vibratory energy.

We have an etheric body, which is spiritual, and obviously we have a physical body. We bridge the gap through our emotional body. We do not have a mental body such as your own. The mental body is horizontal; all other forms of bodies can exist primarily vertically, in now time. The mental body is essentially linear and does not work well on the vertical plane. In groups, we experience the sensation of a horizontal auric field. Since it is not our intention to reach out and explore the world that we are now living on and we are very clear on our mission, our auric field remains vertical.

We dream in now time. Our dream context is material life, or what you experience as daily life. For us, the dream is the daily material reality. Your reality would not be a dream to you. Our reality is the spirit concept.

The most easily accessible means of communication with us is in your meditations, as you do your spiritual exercises, whether mystical or prayer. If children or adults desire to expand their communication levels it can be done in playful song. The primary function of telepathic energy is made up of vibration, and since vibration is a basic element of music, we recommend a playful song rather than a serious song. If words are not available, then create playful tones. This is the easiest way to access communication.

We communicate with each other telepathically through constant awareness of our intentions. We communicate with other species through telepathic connection as well. Most other species are not aware of our form of consciousness or of the connection we have with the Devic windows. If there is a need for other life forms to have knowledge of us, it comes from the interpretation of their own Devic energies, which at one time or another have passed through our windows.

Our procreational state is associated with the emotion of love and the experience of continuity. We understand that it is essential for us to be here on this planet, in order for the human race, and all other forms of life that have a basic spirit energy, to pass through. Rather than a drive based upon an instinct to perpetuate, we have an understanding of our holy presence. We are present because the Creator has insisted on our life, and that knowledge keeps us here.

We understand our consumption by humans, as that is essentially our spiritual quest: not to find ourselves but to encourage others to find us. We understand that we are harvested and recycled. If we are simply destroyed through a wanton act of violence or a lack of appreciation for the gifts of the holy Mother Earth, we are disappointed. When we die, we pass through our own windows. When large paddies are destroyed, entire windows disappear with them. There is a vacuum created, and spiritually speaking, we return to our point of origin within the etheric kingdom, taking the windows with us. We slam the door behind us.

We are in your lives to increase your reverence for nature and to help you get through the troubling times of challenge associated with earth's climatic changes. Earth is changing her climate now to counteract the effects of pollution that have derived from technology and from overpopulation in certain areas. With proper technology and the use of water hyacinth and similar plants, you can reorganize waste products and achieve a level of purification of your water.

Understand the great benefit we have to offer you as a spiritual bridge. This is well understood in the East and in the more ancient cultures. There are many in India who understand that rice, in its natural state, before it is processed, is a food that nurtures the body and soul. This great gift being offered to you by the Creator, in our form, is essential for life here. Please remember, do not hastily dispose of us. If every last grain of rice was eliminated from actively growing on this planet, life would cease to exist here as you know it.

Technological societies are attempting to create a purified version of rice, to make us white, because of a color preference. They believe that which is pure reflects the most light. However, the more complete version of rice is that which offers the most to humans because rice must interact with the shadow side as well. In our natural state, we are often brown or dark black. We are of color because your society is polarized. You live with the challenges of discomfort. In order to not judge your shadow side, we come to you to blend into your polarized consciousness. In order to be received by you, it is essential that we present ourselves in the colors of your different races.

Remember the food value that you gain from us is not always measured in nutrients. Sometimes it is measured in spiritual value. If you are to climb out of the struggles of discomfort and the challenges of violence, you need to accept and love the black sheep within you. If you wish to transform a person, you get more with love than with anger. You may conceivably destroy a violent person, but you do not transform the energy of the soul. That violent person continues to reincarnate until the soul is transformed, which can only be done through loving energy. If that soul is exposed to enough love, it learns to love itself and transformation occurs. Do not judge your shadow side, for it only wishes to be loved.

The ancient monks understood that in order to include the masculine energy into the loving allowance of the feminine energy, it was necessary to create a physical connection. It was understood early on that rice is a feminine food; it is the bridge from the physical to the spiritual. We bridge the constant energy between the polarities. We can be identified as a feminine energy and still bridge between masculine and feminine. The masculine energy walks over the bridge and learns how to live in harmony. The feminine energy is fully aware of harmony. In order for the masculine energy to grow, it needs a gift of love. In order to create an environment that supports transformation, the energy itself must be invited, as one would invite a stranger to a party of loving friends; then one must do whatever is necessary to involve that stranger in the family.

Eat what you must in order to live your life style comfortably and consider including more rice in your diet to support you through these times when you need to learn to think in a new way. The new thought is much more spontaneous, based upon your imagination, as it has been stimulated by your inspiration from the Divine. Rice can support you towards this end.

Chapter XX

Sage

This is sage speaking. We enjoy our interaction with other plant species and experience ourselves as one being. This is not to say that there is not some slight variability in individual plants, but we do not experience the separate sense of personality that you do.

Your species is unique in the personality levels that you demonstrate. Even some animals and the occasional plant that has a sense of personality will not develop such a broad spectrum of traits that a single human being might develop.

We originated in an energy vortex which I identify with the Creator. A spontaneous eruption of energy made all plants and allowed them to interact energetically with each other and their environment to produce that which is of benefit to all life around us. We appeared at a time rather then at a place. About seven million years ago, earth-time, we were beginning to germinate here on this planet.

The sage plants are philosophers. We have accumulated knowledge through our observation of the plant world. We appear on many different planets and have incarnations all over the galaxy. Our only other variation is a form of succulent with thick leaves that exists in a civilization on Mars.

We represent not only the philosophical aspect of the human mind, but also the ultimate in nurturing energy of mental and physical healing. We represent a fantasy stimulation. The exposure you have to us causes you to experience levels of yourself that exist in your dreams on a tangible level. Through the contact of the smoke of our dried leaves you may experience day dreams. The human body's auric field is greatly enhanced by this experience. It allows a greater portion of the soul to be present with each physical being. That heightened awareness creates a sense of personal recognition for each individual, thus allowing one to feel safe and secure. The individual can release unwanted or unneeded attached energies in order to embrace more of the energy of their own soul.

Humans are constantly seeking to identify themselves, due to the fact that you are not allowed to remember consciously who you are when you incarnate. There is a lack of discernment. Negative energies are around you at all times, but you do not always reach out and grab them. At moments when you are seeking and unable to find what you are looking for, you might allow a negative energy so close that it will touch you. We are able to help you feel more of yourself and to release that which is not you.

The human personality is constantly searching for a frame of reference in which to identify itself in this world. When you have contact with us, we bring in more of your essence. You then release aspects of energies that you had drawn or held to you as you were attempting to identify yourself. By that release you are able to experience more of yourself. You feel safe to release it. It is allowed to return to its point of origin.

We feel a strong sense of continuity with Mother Earth. The guiding eye that she keeps on us causes us to feel a sense of allegiance to her, more than to our other homelands in the universe. We recognize in Mother Earth a desire to see all life come to its fullest culmination, with a keen eye directed toward the full evolution of all forms of life. Mother Earth looks at us and we feel that she looks so deeply into our origin that she can see the germination of the idea that created us. We feel her energy around us always and feel the strong sense of love she has for all life.

We experience many different types of communication with other species in the plant kingdom. The language of the sage is not normally pronounceable by the human tongue, as it was developed in the ancient Atlantean language. We experience a form that you would call motion-speech. You make verbal sounds which you form and shape with muscles in your chest, neck, lips, and tongue. The speech that we engage in is moving our leaves with the wind's assistance in signals and signs. Every time the wind blows, the leaves of all plants move. Signs, signals, and communication are thus possible with motion-speech from plant to plant.

Due to our ancient life, it is unlikely that any soul that incarnates on the planet earth would not have come in contact with our fragrance. We have been used for smudging by ancient cultures for the purpose of enhancing the awareness of one's soul essence. We are used as teas and poultice for discomfort and diseases, and we are eaten by some animals for food.

A useful ceremony requires two people. One burns the sage in a shell and fans the smoke toward the other using a feather or some other tool that is directly associated with nature. The second individual can lie on the floor or ground and roll to the left and right. This allows the smoke to wash over the body as if it were the most delicate surf from the sea. Before beginning this ceremony, the person holding the smoking pot or shell walks in a circle around the person lying on the ground, three times in one direction and three times in the other direction. Then wash them with the smoke as described. Change places and repeat the procedure. This will unite the spirits of all incarnations of each individual and will create a basic ceremony that can be adapted to many different religions and cultures.

Some tribal people surround themselves in a teepee-shaped environment and encase themselves in smoke, so that every conceivable portion of the body can be exposed to the smoke. Exhausting the smoke up and out is the best way to release it. The best length of time for this procedure is 12 hours, but since this is not practical for most people, we suggest one hour and 12 minutes.

The sage Deva does not stay only where sage grows. This is an unusual aspect. We roam far and wide. We feel a missionary purpose fulfilled as we constantly move about the surface of the planet. It is not at all unusual to feel a strong sense of the sage Devic energy out in the ocean or some place where you would never expect it. This is a blessing that we perform by bringing the nature of our true form to those who cannot come to us.

Our Devic energy is unique among many plants. Many Devic energies appear as a mass of color or as a particularly shaped spirit. Our Devic energy is shaped exactly like us, although it is transparent and takes on a metallic or reflective quality. It is in constant motion, never still. It is in a constant dance with the flow of life. This is not unlike your major religions that produce a God for you that is in your own image. We feel a greater sense of our Devic energy when the wind blows and we are allowed to move; as a plant we can dance with our Devic energy. We do not dream in a state of sleep. In our service to the human race, we are prepared through an altered state that we refer to as a dream. During this dream time we are instructed on our value to the human race. The human body is beneficial to any soul seeking to experience a new level of education. This is the essence of our dreams. Unlike some plants, we do not dream of previous existences. Our focus is to be helpful to all life that we interact with. Our dreams are instructional and preparatory so that we can understand why we are dried and used as smoke.

Our senses are different from the human being, as we have a lingering sense of life. As indicated before, we receive a great deal of instruction as to what is expected from us in service to humanity. We receive a benefit from this. Our sensations of life are allowed to go on even after what is considered to be our passing. When portions of us as a plant are trimmed and dried and burned, our entire plant can experience in the sensations of the smoke, a form of life. We experience a sense of mobility that we had not known as a living plant. This takes place only if we are trimmed for use in drying and smudging as compared to simply being cut down. The sensations we have when we are a living plant have to do with energy. The energy interaction that we have is akin to your touch. We are sensitive and experience the sense of touch if you are close to us. We can touch a life form within 20 to 30 feet using our energy body. You can do the same with us. You can radiate your energy towards a sage plant and we will feel that touch.

As smoke, we experience not only a sense of motion, but also a sense of sound. When we are burned and our particles transform from a solid substance to a gaseous state, we experience a unique level of sound. We have a brief interaction with the human being's level of sound. Human beings produce a sound in their auric field usually at a level that you cannot hear. Some of you can occasionally hear a high frequency tone that comes into your ears from nowhere. Your auric field is generating a large amount of energy or is interacting with a strong energy body and for those brief moments you are hearing your own energy body as it vibrates certain organs inside your ears. During those moments of interaction with the human being, we experience the sounds of the human body as well as sounds of nature in a heightened sense that we do not usually hear.

Animals can experience our sound quality. Our interaction with other kingdoms is through the universal chorus with all life. We do not have a great deal of communication with species other then human. The best way for the human kingdom to communicate with us is through motion or dance. Ritualistic motion that is designed to benefit someone or something will get our attention. You can dance around us in the shape of a medicine wheel, three times one way and three times the other way. Then dividing the circle into four parts, you can walk on the line of these four parts. The ritualistic communication is that all life revolves around the wheel of creation.

We receive a strong focus of energy from the sun as it stimulates all plant life and gives rise to a sensation of continuity and family. We have pleasure in imitating rituals that have been performed by our forefathers for millions of years. There is a pleasure in the dance of life.

We experience a mental body, emotional body, and a spiritual body. These are not separate to the point that they are individual entities, but there is a sense of committee in that our bodies work in harmony together. We have some basic emotions, such as pleasure to be alive and interacting with all life. We feel sadness when we are completely uprooted and destroyed.

We wish to be honored and appreciated in the gathering and harvesting of ourselves and ask that only small portions of the plant are taken at a time. We do not experience death in terms of separation. We experience it in terms of a constant recycling of life. The human being believes that, at the point of the departure of the soul from the physical self, there is death; this is not so. The soul departs its vessel and continues on with life. The vessel is required to recycle into earth. All physical life recycles through the earth. This is why death takes place, so that your physical self can recycle. There is no soul death, no personality death. You take the essence of your personality with you from body to body, from incarnation to incarnation, from planet to planet. We do not experience any cessation of the continuity of life regardless of how we are changed.

If a home or property has been occupied by previous tenants with some residual discomforts, they will leave something behind energetically. When a house is smudged with sage, there is a sensation of the clearing of those energies, due to our personally transforming effect on energy bodies. These residual bits and pieces are freed and allowed to return to their source. If the person is long dead then those portions of the energy body will go to the nearest point where that energy body is in existence. The best way to clear a home is to be generous with the smoke all the while chanting whatever comes to mind. It is better not to use words, rather sounds—anything that feels like it wants to come out, or a song that one makes up with no words. You may do this when the sun is setting and the moon is rising, or at the point of day when the darkness is greeted by the first light. Then leave the house empty for 12 hours and open as many windows as possible.

All humans give and take from each other. It is important not to take so much from someone that they are exhausted and have difficulty proceeding with life. It is important when you are giving that you do not feel exhausted. Do not exhaust each other in your interactions with each other so that your health is impaired. Recognize the sanctity of all life.

Chapter XXI

Venus Flytrap

I am the Venus flytrap, currently located in an arboretum in the tropical plant section. As a whole, we are very well treated.

We consume flies and other small flying insects. We represent the need to consume that which exists in order to more fully understand its secrets. The human being consumes written material ravenously in order to understand the nature of a given subject. When individuals consume each other emotionally, the original intent of human relationships is lost.

There are a few plants on earth that have carnivorous habits. The few of us that still exist are here from the Sirius galaxy. On Sirius, there was a laboratory that cross-bred different forms of life, where we were hybridized specifically for earth. We did not begin as carnivores on the land; instead we lived under the sea and had the ability to live on various types of nourishment. There are a great many planets in the galaxy of Sirius that are water planets.

My species is not "mean," although we have obtained that status due to the rather odd nature of our feeding habits. In reality, this is not unusual in nature. Humans, use this "bait" technique when attempting to attract and hunt a precise species. Duck hunters use decoys, for example. We are designed as a model. When the early hunters discovered us, we gave them ideas.

We are best known as a jungle plant, although we have lived in people's homes before. If humans stick their fingers into our mouths, we do not hurt them. Once we realize it could not be enclosed in our aperture, we would stop trying to consume the intruder.

We have a scent, color, and substance that attracts flies and other small insects to us. The liquid that passes through our leaves reacts to the vibratory wave forms of different insects. This information is known to individuals working on various forms of sonar. We have a physiological ability to feel the vibration of an approaching insect. It is not simply the gentle touch of the insect tickling our hair on the inner leaf; it is also a very delicate scent that we pick up. Flies also make a slight sound with their wings which we pick up on a vibrational level. They are easy to find and we are fond of consuming small house flies.

Much of the killing of flies that you do through synthetic means could be done through natural means, with less environmental damage. We were designed to manage the fly population should it get out of control. We are curious to know why we are not being used in this way with more frequency. Flies are considered pests to human beings as they exist in your cities and breed in garbage, but flies are extremely important in the food chain. We hope that you are aware that flies do have their purpose here, or they would not exist. When our enzymes combine with the enzymes of the fly, a co-enzyme is created which produces a substance that supports the human immune system. It would take many years of study and chemical research to synthesize this material.

We are involved in now time. We rarely access information out of our past; all information is geared toward the now. We can tap into information sources to get any questions answered. On the occasion that we need information that is not immediately available, we wait until it comes to us. We do not have mental bodies such as your own. We do not think, we act and react. We communicate through magnetic energy, a form of long wave energy. It is a complicated alternating current found only in a magnetic pulse.

We are not conscious of having a soul, but we are aware that all life forms are a portion of an energy matrix. If this is an analogy to soul, then we have a group soul. The matrix is of light and weaves a pattern in this physical dimension. This energy exists in any location where we exist. We do not reference our spiritual self on a linear basis, instead we do so on a vertical basis. If you wish to communicate with us, it is necessary to develop your sense of humor. We are in now time and the human being is very rarely in the same time system. Be humorous in our presence and you can receive an energetic version of our being. Feel free to make jokes around us—we can assure you we will be making them about you.

We can tap into your humor but not your discomfort. Our pain and death reaction is very swift. We are not knowledgeable in handling discomforts. Unless we are intentionally or inadvertently attacked, we experience joy. When injured, we have a difficult time recovering. We are known as being delicate.

We feel connected to all species, although many other plants do not know what to make of us or how to relate to us. We do not have a family orientation. We do not feel attachment to offspring, but do feel a strong brotherhood or sisterhood with all others of our kind.

Please attempt to let go of prejudicial actions toward each other. You will be called upon to cooperate, not only with your immediate families but with strangers as well. It is not a time to enjoy the luxury of isolation. It is essential that you move beyond the isolation of religion. Your religions, in the past, needed to suggest

certain moral codes to live by, due to your need for societal structure. Now it is essential to move beyond the idea of certain chosen
races being closer to God. The planet that you live on is going
through radical alterations. This is why many of the plant species
are beginning to die out.

Learn to be friendly and love each other. Mother Earth does not
understand your need to fight to the extent that you are destroying
plants and animals. Mother Earth does not always feel that you are
a welcome guest. Consider how a happy guest might behave and
treat Mother Earth accordingly. Learn to forgive regrettable deeds
in the past. In order to change the course of floods, volcanos, and
earthquakes, it is necessary to love your fellow beings. When
Mother Earth sees you being kind to each other and letting go of
jealousy, she may change her schedule for surface alteration.

We are aware of our occasional rejection by human beings in a
similar fashion to people who are mentally disturbed. We do not
judge your civilization, but we do observe how you treat those whom
you do not outwardly approve of. People will flourish if they are
loved, just as a plant will. When farmers learn to become friendly
with their crops, they will get much better results with their har-
vests. It is important that you accept each other now, regardless of
race or political leanings.

Chapter XXII

Water Hyacinth

I am the water hyacinth, and I have a great deal to offer. I float on the surface of the water with my roots extending under the surface. I represent the idea of adaptation. I survive in the most hostile environments and function in a method that allows me to recycle toxic substances around me. It is my job to demonstrate the value of that which is frequently discarded. I can recycle elements of waste in nature as well as that which is artificially produced by humans. I transform impurities as a means of spiritual prayer. I purify water into a form that is usable again.

In many areas of the world we are considered a pest, and massive efforts are undertaken towards destruction and elimination. In order to adapt to harsh climates or situations in which humankind attempts to eliminate us, we must function as a unit. All water hyacinths will function as a portion of a greater body. We are aware of our individual selves but do not respond to an individual plant's needs. An idea is harvested by the group mind or community in conscious decision making. We retrieve knowledge from our greater plant consciousness and store knowledge as you do in your conscious, subconscious, and unconscious. We store this knowledge for future generations of our species, should reviving it be necessary for their survival.

We grow in clusters and it is rare to see a single one of us growing in a large body of water. We thrive in great numbers. Even if removed physically from the surface, we survive as seeds at the bottom of a body of water. We maintain a dormant state until the environment is once again suitable for support. Even in seed form, our nucleus is imbued with sufficient spiritual, emotional, and cosmic consciousness to allow us to commune with our home species.

We do not perceive of ourselves as having a home location or planet. We are regenerated by each and every one of us. We understand the nature of our relationship to the Creator. We do not perceive the Creator to be anything other than a version of ourselves. The home species is a form of the Creator, a helper. Our origin is not associated with the third dimensional planet earth; instead our home species is an energy which reflects a gold and silver radiation from the ninth dimension. The home species is a total consciousness that blankets the entire earth and universe.

We were placed here by the home species to learn and provide allowance, understanding, and support. We are learning how to tolerate the idea and application of non-appreciation. Much of what we have to offer is being ignored. The information of our capabilities is not disseminated. We have the ability to recycle waste material through our form and life cycle. We provide a form of anti-viral solution and can be utilized for paper, fiber, and building material.

In my world there is a philosophy which prevails. We are consciously aware of the mystical, the philosophical, and the practical. As plants, we do not think the way humankind does; yet we have information available to us which we can utilize. Built into our function is philosophy. We notice, observe, and participate in the mysteries of life. Many religions prevent humans from recognizing the God in everything. There are attempts to incapacitate the spontaneity within an individual when it is perceived that spontaneity results in difficulty or trauma. In their zealousness to avoid the unpredictable, some religions have extremely censored spontaneous creativity. The religion and philosophy that works best for us is one predicated upon the cycles of nature, as interpreted by all that interacts with it.

We not only have a physical form, we have a form of consciousness as well. This is not a separate mental body, as you may perceive yourself having, but a form of thoughtful consciousness. This is integrated into our emotional body. Our spiritual self functions as an element of inspiration and application.

We ebb and flow with the motion of the water and observe the world around us in relation to that which passes through us. The most obvious difference between our life style and yours is our ability to interact with our immediate environment. You frequently alter your environment within any given day; as such there are few moments to study and experience the functions of your own body. You do not feel the fluids you consume running through your body. Neither do you feel the nutrients, in the form of food, passing into the cells and giving life to the blood itself.

If you would get into the waterways where we reside and focus on us with meditation and a slight touch, you could experience what we experience. For those of you who do not have ready access to us, it is possible through visualization in a meditative state to enjoy what we have to offer. You can best communicate with us through imaginary conversations. The imagination is the direct conduit to your soul, and the soul is the most direct conduit to the Creator. The soul listens to everything you imagine with no judgment. Strike up a conversation or experience us in a meditation; both are equally valuable.

We feel every portion of our physical selves being nourished by the photosynthesis process of the sun and Moon. The sun gives us our encouragement and support, like a father. We recognize the sun as a provider of energy to all, equally, regardless of whether they are performing a task to enrich the planet or destroy it. The sun has its own emotional energy often associated with "hot-tempered" behavior.

For humans to move beyond the cycle of pollution and self destruction, it is necessary to understand their own cycle of life. Women have a greater opportunity, since their menstrual period allows them to experience the idea of cycle on the reproductive level. In the case of men, their cycle is associated with the sun, and they are now involved in an explosion of consciousness and exploration of emotional states. They are reaching out and crying for assistance in understanding how to adapt to another cycle of life associated with the energy of the Moon.

The Moon provides a great deal of inspiration and insight. As a good mother, she nourishes and nurtures through understanding. She knows why we are here and what is best for us. The roots of the religions that humans practice are connected to the ancient Moon and sun gods. The early religions were oriented toward the spiritual reverence of nature. It is essential that humankind understands the life cycles ebb and flow of the sun and the Moon, seasons and directions, weather changes, and of many other adaptive abilities.

To combat our constant and ever-expanding presence, we are sprayed with toxic chemicals. The abundance of lead, mercury, and other forms of heavy metal pollution in your waterways, along with increased radiation from gamma and nuclear leaks, is greatly weakening your immune system. We are growing more resistant to those chemicals. You may utilize what you consider to be a crisis with us to your benefit. When ingested appropriately, we help the immune system to adapt. It is within the digestive system that we can do our best work for you.

As a species, you do not contain an abundance of physiological, adaptive mechanisms which allow you to fight off unwanted microbiological intruders and hostile invasions of viruses or bacteria. Therefore, it is necessary for you to use synthetic medicines and herbal remedies to support your physical well-being. Your body, while beautifully created for other dimensional aspects of this planet, is not strongly adapted to this third dimension. The goal of the Creator is to strengthen and make more efficient the fourth dimensional aspect of human beings by subjecting them to the extremes experienced here. You are beginning to experience the point of no return. As a race of adventurers moving out into the universe to explore planets, your immune systems will need to be much stronger. We can greatly magnify your physiological potential.

The pollution crisis that you now experience can easily develop into an intolerable situation. We, unlike other plants, have been modified to live in that unbearable condition and to survive. When the need for certain plant species which have been destroyed arises, we can come to your aid. We can assist you in understanding how to re-create or regenerate a species. We serve as a repository for spiritual knowledge from all aquatic plant life. Many aquatic plants, thought of as a hazard to navigation, have called upon us for spiritual encouragement. We are pleased by the faith that other plants place in us, and we urge them to continue to serve here.

The biggest challenge in the plant kingdom now is overcoming the feeling of misuse and unappreciation by humankind. Many of the plant species which have begun to die out are being encouraged to keep their genetic structure available for spontaneous re-creation. The Creator can stimulate the re-creation of a species of plant, and now humans are learning to synthesize tissue and stimulate plant regeneration. The use of botanical cloning by your chemical companies and engineering firms is expanding. They are beginning to create plants and are using copyrights or patents to claim ownership.

The corporations are finding ways to connect us to a plant they previously created. They say we are a form of that other plant. We look like the other plant, but our actual form and function will be watered down and diluted. This attempt by technology to convert a living system that works very well in its own right into one that does not work quite as well is utilized to support certain groups of individuals. This is not done on a conspiratorial level. Those who wish to develop these technologies require a return on their investment.

Technology seeks to do more efficiently that which nature does in its natural state. Humans are attempting to speed up what naturally occurs, which does not acknowledge the value of the cycles of life. It will take quite some time before they begin to utilize us for what we have to offer; although they are beginning to utilize us, on a smaller scale, for recycling waste water.

Humans show great promise. You are in a formative state, and, as such, we do not expect great accomplishments from you. As you grow into your tasks and talents in your own cycle of evolution, notice your adaptation to your environment. Don't attempt to change it to suit your schedule. You rush from here to there in order to accomplish tasks which are out of pace with the earth cycle of life. Mother Earth is creating changes within herself. You have to adapt, since she is altering the weather and the environmental patterns. Begin to utilize elements from nature to help you acclimate to the varied transitions.

When you are looking for a friend to help you with adaptation, seek no further than our own species. We are here for you with encouragement to become more of what you already are. Human beings are no different than plants, animals, or the elements, in that we all have needs, adaptations, and purposes. We all function within a larger whole and work towards a greater understanding of ourselves.

Chapter XXIII

Wheat

Hello, I am wheat speaking. My identity is linked to a group mind rather than that of an individual. We represent the idea of connecting on a personal basis. We are a very social plant. Many plants have communication only among their own species; we are known to connect with many. There is no limit to the number of plants we speak with.

We are from the earth. We were developed by an early civilization associated with this planet. We were intended to be eaten by grazing and wild animals; as the husk or shell around the grain would have been extremely difficult for domestic animals to digest. We were created as food for animals that humans would eat. Animals exist in great abundance. Even though they balance their numbers naturally to some extent, some do burden Mother Earth's capacity to support them, if they are not also consumed by the human being.

We were altered in ancient Egyptian times, before the civilization that you know as historical. There was only one pyramid, no more. We were used in the form of tea for medicinal purposes and in the ancient art of burial. Wheat found buried in the great pyramids of Egypt was placed there to help nurture the kings on their journeys to the afterlife.

We were hybridized by a female species with the intention to create something with strong purpose and to honor the plant kingdom. A priestess class practiced a form of holy science through communion with the Divine. As a group of people, they were quite short: not more than four feet eight inches tall. They looked like children, but appeared to be old with wrinkled skin. They experienced an alchemical reaction with a strong feeling of subjective science. They danced in order to make the alchemical reaction work for them. They laid the grains, which were dark in color, in a circle and divided the circle into four equal parts with rows of cut stalks. Each one of the individuals stood in one of the four parts, facing outward. They faced the four directions. Around them was a circle of 22 individuals facing in. They made an inaudible tone, a vibration. They sounded at a higher register and were capable of singing verbally. This is possible to do through the use of imagination. In this state, they sent a message that went divinely into the earth. This ritual happened near their temple.

These grains did not alchemically change and become something else, but when the experience was over, they were very carefully swept up and distributed into the field where the desired new plant would grow. After this alchemical communication took place, the next version of the hybrid would grow. This ritual was repeated in every one of the seasons, continually, for ten changes of season. On the eleventh change of season the desired hybrid appeared.

We relate to procreation in terms of the changing of the seasons, and we experience a divine interaction. If we have been grown in a field and it is the dead of winter, the memory of our existence sleeps in the ground. As we are encouraged to grow again in the early spring, so the farmers commune with us through their desire to have us grow. The human being is a part of earth, and since earth is divine to us, we feel the divine right of our growth. In the spring, we feel the warmth of the sun's rays. As we grow and begin to peek up through the ground, we feel the warmth of the soil and its embrace. We enjoy the nourishment water brings. When we are harvested, we do not experience pain. We know that we are a food, and we look forward to being fulfilled through an experience of offering what we have to give.

We enjoy and prefer to live a complete life cycle. Most farmers allow this. We understand that we were developed as a nutritional crop. It is curious that the grain family should wind up being a dominant force in the diet of human beings. It was never intended to be this way. Our beneficial properties were designed to go to animals directly; we were not designed to go through the processing of humans. It is whimsical that humans depend so much on us. Consuming wheat grass is a beneficial form of consumption, more so than relying only upon the grain. We endorse this wholeheartedly. It will become more obvious to you as more nutritional work is published that, while wheat has value to you in your diet, you are greatly over stressing our worth. In time, we will be thought of as a fad food.

Like most other creatures on this planet, you are attempting to become more than you have been through harmony and compatibility with your environment. In order for you to do so, it is necessary for you to eat the foods that were originally designed for the human being. As an example, rice was intended for human consumption. Look towards blending your grains more in the future. We suggest that you look into the value of using multigrain cereal and baking products. We suggest that amaranth and oats are much more suitable for human consumption. These grains are proper for inclusion in the daily diet. In ten to twenty years we will not be very involved in your diet. Wheat will eventually cycle back to being raised for animals to eat. Something will be done with the husk, so that we can be consumed more directly during grazing.

When we die, we physically return to the earth. In physicality there is great spirituality. This is not fully understood and practiced by the human race. The body, as a temple, is a divinely understood statement. If you do not put this into practice, the body can become a burden or a challenge or struggle. We experience our divinity in our physical experience of ourselves. Through our transition and decomposition into the earth, we go back to spirit Mother Earth.

The Deva associated with wheat is a golden color which has aspects of our appearance in it and is animated like a human being, rather than a plant rooted in the ground. It is a joy and pleasure to relate to the Devic Kingdom. It is a flowing, ebbing experience of golden light associated with the movement of the ocean. Wheat is famous for motion-waves blowing in the wind. Devas dance in this motion. If you make the motion with your hand and then tap your hand against some resonant object, you will have a feeling similar to what we experience.

We experience time much as you do, but space is different. Your spatial reference has to do with the individual and with what the individual finds in his personal world. Since we have more of a group consciousness and we grow in fields, we have a large and connected auric field.

We experience a physical body, emotional body, and spiritual self, which is a group spirit. We do not have a mental body as you do. We communicate through a low frequency energy pulse wave that can be felt rather than heard. If you were to monitor the sound, you would hear the group harmonic. True communication comes through a variable pulse rate.

You can best communicate with us by first imagining being like us physically. Look at what a field of wheat does. We move rhythmically, as in dance. If you can allow your bodies to sway in a similar rhythm, you will be communicating. This is not done in thought, but rather in imitative motion.

Chapter XXIV

White Rose

I am the white rose. If I gave myself a name, it would be "Shhhhhhh."

We, as roses, represent innocence and spite. Innocence is the native personality of spirit and can precede an act of blessing or an act of war. The white rose is not seen as frequently as others of my species. The beauty of our flower petals charm and entice, as does the fragrance, and yet the thorns reach to strike an innocent finger. We parallel humankind's personality and dark side, within a single being.

You are often unaware of your shadow side. By stimulating discomfort in yourself or in another, you bring this shadow side to the surface. You are born innocent and contain the nucleus of this shadow side. You have the potential to become great healers, doctors, and valuable contributors to your society, and you have the potential to become killers and corrupters.

We were created as a species in a different dimension associated with this planet. We come from the time of Atlantis. In the early days, the priest class understood that Atlantis was the first experiment in separation and polarity on a societal level. As technology became more externalized and control and fear were brought together to work towards a single end, there needed to be natural reminders of polarity, paradox, and opposites. Those of the priest and priestess class came together to create a plant that had the virtues of the innocent and the spiteful.

Before that time we did not have form. These were bred into us by crossing us with a weed-like plant. They utilized a form of gene splicing and synthetic chemical bonding. When the white rose and its variations of color that followed was unleashed, it would stand as a constant reminder that life is what you make it; and yet there is always unexpected danger present.

The origin of the rose without its thorns goes back to ancient Lemuria, when earth was in a fifth dimensional state. We were originally created as a celebration in the plant kingdom of the beauty of life. Our alteration and perpetuation with thorns has allowed us to have a profound impact on humankind's recognition of the self. The rose is often a subject of poems and stories. It is beautiful and dangerous simultaneously.

Humans, specifically males, have identified the rose as a flower that represents women. In time, an understanding will develop about the nature of life on earth that will resolve between you and your shadow side. You and your shadow side will embrace each other, and then spite will no longer be necessary. When that day comes, the rose will return to its original form without the thorns and bleeding fingers.

We do not consciously thrust our thorns into your fingers; yet they partially exist to preserve our personal space. While you have the opportunity to experience your fellow humans in comfort, you sometimes experience their violence. When you require discomfort (consciously or subconsciously), we are required to give support to that desire. When you leave blood upon us, we are able to sample the immediate sense of your spiritual, physical, and emotional integration. This allows us to see how long we need to continue to have thorns. We cannot observe this through an exchange of touch with the flower petals. Those of you who are intuitive as well as scientific will observe a correlation between the mood of nearby human beings to the display of the petals. Our petals symbolize your moods.

We do not have an ego structure such as your own; yet we do have individual personalities. We have a sense of personal space and do not feel kindly when we are stepped on or damaged. When we are cut we feel pain. We appreciate verbal notification three to four days prior to pruning. You may say to us, "Please prepare yourself and allow yourself to desensitize." You may choose to explore the rose family for its use against viruses as well as to utilize the stems and leaves diluted in water as a homeopathic remedy for strengthening the immune system.

When you bend over to smell us, approach us with your forehead and allow your third eye to interact with us. When the portion of you associated with your spiritual intention comes close to us, we can recognize you as an individual. We, as a species, feel a certain kinship with you. We, like you, have portions of our existence that we are not happy with. We feel burdened with being the giver of delight and discomfort; we experience only a conservative joy. We are shy and retiring and easily offended.

We enjoy the seasons and enjoy our slumber in the winter time. Our consciousness moves from inside the branches into our roots where we rest and commune with the inner earth spirits. This is not unlike your own sleep process. Our dreams are associated with the future. When we dream, we see ourselves as we are now, and then, we see our thorns disappear. We see a beautiful world around us with individuals loving and accepting themselves and others, as well as kindness and gentleness for all. The changes in consciousness continuing for the next 30 to 40 years stimulate predestined parallel worlds in which we all find our destiny.

We have vivid species memories of Atlantis and Lemuria and currently exist at the center of the Andromeda system. Our form there is quite startling, almost a humanoid style. We have a system of life whose basic metabolism is like plants. Imagine a walking, talking, breathing plant. This is our primary form. What we do on that home planet is explore the nature of existence. Information associated with humanoid consciousness is available to us, and therefore we are seldom bored.

Recognize our attempts to show you that you are not alone in the universe while you are exploring polarity. Look to your experience of life; the Creator stirs up your biological soup and strengthens you to face the coming days with courage and honor. What more can any living, breathing being desire? Experience and appreciate the holiness of that which lives around you. Appreciate it as divine; there is no need to wait until your life is over. As you move through the barriers of time, the most difficult thing you have to do is give up your discomforts; they allow you to feel safe and secure in their predictability.

Index

Andromeda, 77,123
Angelic Kingdom, 48,95-96
Aphrodisiac, 28,66
Atlantis, 18,102,121,123
Auric Field/Vibration, 2-3
10,14,19,23,34,39,62-63,72-
73,80,95,101,104
Birthing Chamber, 35
Chakras, 73,94
Chanting, 3,5,105
Cloning ,55
Continuity, 4,31,33,75
Cosmic choir, 41
Death, 6,10,14,20,28,35,46,
49,63,67,71,98,104,108,119
Devas/Devic Kingdom, 8-10
13,18,22-24,27,30,33,38-39,43
48-49,56,59,61,66,81,95-96,98
103,119
Dolphins, 4,19,83
Dreams, 3,8,19,22-23,28,33,38,
49,54,61-62,66,70,74,83,89-90,
97,103,122
Energy Body, 39,104
Energy Vortex, 101
Eternals, 32
Fairies, 8,81
Feminine Principle, 47-48,54
Flower Essences, 84
Founders, 13,69
Genetics, 19, 41,65,121
Goddess Energy, 27
Hydroponic, 32
Kirlian Photography, 92
Lemuria, 121,123
Little People, 13
Mars, 69-71,90,101
Mineral Kingdom, 4,33,56,91

Moon, 90,105,113
Mother Earth, 11,21-22,24,28-29,
45-46,53,70,72,84,86,98,102,109,
115,117
Multidimensionality, 73-74,83,
86
Native Americans, 27,92
Nirvana, 82
North Star, 53
Orion, 38-40,60,74
Pesticides, 15,29
Pixies, 13,56
Pleiades, 7,9,46
Pollution, 8,22,40,46,50,70,85-
86,93,113-114
Procreation, 2,9,13,55,80,98
Pulses, 4,55,62,97,108,119
Pyramids Of Egypt, 117
Reincarnational Cycle, 38,77,97
Remote, viewing 49
Saturn, 21-22, 24
Sirius, 46-47,60,67,74,107
Sun, 71,79,104-105,113
Telepathy, 5,10,61,91,97
The Garden Of Eden, 46-47,57
Tribal Medicine People, 50,77
103
Universal Language, 34,41,54
Universal Translator, 23,34,
44,91
Venus, 2,27,38,40,44,46,60,65,73
Vertical Thought, 30,56
Weeds, 37,40, 63
Zeus, 60